SIMPLY IN DEPTH

THE INTERNET OF THINGS

IoT

AJIT SINGH
SULTAN AHMAD

Preface

The "IoT Simply In Depth" book is in active development by a joint effort from both academia and industrial collaborators, acknowledging that the Internet of Things of the future will be built on top of scalable and mature protocols, such as IPv6, 6LoWPAN and IEEE 802.15.4.

The Internet of Things can be characterized as joining the physical object, the computer embedded into it, and communication and code on the Internet itself. We focus on these three elements in both the prototyping and the manufacturing sections. We began by looking at some examples of the Internet of Things in action. Throughout the book, we discuss many REAL LIFE projects, We have tried to cover introduction, implementation of IoT using Arduino and RASPBERRY PI along with suitable Case Studies.

In this book we look at the kinds of computer chips that can be embedded in objects ("microcontrollers" such as the Arduino) and take you through each step of the process from prototyping a Thing to manufacturing and selling it. We explore the platforms you can use to develop the hardware and software.

I'd like to thank Sultan for agreeing to join me in the adventure of actually doing the writing. We have consistently reviewed each other's contents before submitting them to KDP, which has helped maintain that shared voice through the whole process.

The acknowledgements section was one I never normally paid a lot of attention to, as it was the author thanking a load of people I didn't know. Having written a book, and realized how much help and support are given to the authors, I have a newfound appreciation for this section.

The Internet of Things

For information about this title or to order other books and/or electronic media, contact the publisher.

Ajit Singh & Sultan Ahmad
ajit_singh24@yahoo.com
https://www.ajitvoice.in

Table of Contents

About Author(s)

Ajit Singh
Assistant Professor (Ad-hoc)
Department of Computer Application
Patna Women's College, Patna, Bihar.

World Record Tittle(s):
1. Online World Record (OWR).
2. Future Kalams Book Of Records.

A PhD candidate at Patliputra University, Bihar, IND working on **"Social Media Predictive Data Analytics"** at the A. N. College Research Centre, Patna, IND. He also holds M.Phil. Degree in Computer Science, and is a Microsoft's MCSE / MCDBA / MCSD.

20 Years of strong teaching experience for Under Graduate and Post Graduate courses of Computer Science across several colleges of Patna University and NIT Patna, Bihar, IND.

[Amazon's Author Profile]
www.amazon.com/author/ajitsingh

[Contact]
URL: http://www.ajitvoice.in
Email: ajit_singh24@yahoo.com
Ph: +91-923-46-11498

[Memberships]
1. InternetSociety (2168607) - Delhi/Trivendrum Chapters
2. IEEE (95539159)
3. International Association of Engineers (IAENG-233408)
4. Eurasia Research STRA-M19371
5. Member – IoT Council
6. ORCID https://orcid.org/0000-0002-6093-3457
7. Python Software Association
8. Data Science Central
9. Non Fiction Authors Association (NFAA-21979)

Sultan Ahmad

Senior Lecturer-CSE
College of Computer Engineering and Sciences
Prince Sattam Bin Abdulaziz University
Al-Kharj, Saudi Arabia.

Sultan Ahmad is currently working as Senior Lecturer in Department of Computer Science, College of Computer Engineering and Sciences, Prince Sattam Bin Abdulaziz University, Al- Kharj, Saudi Arabia.

He has a unique blend of education and experience. He has received his Master of Computer Science and Applications from the prestigious Aligarh Muslim University, India with distinction marks. He has graduated in Computer Science and Application in 2002 from Patna University, India.

His research and teaching interests include Cloud Computing, Fog/Edge Computing and Internet of Things. He has presented and published his research papers in many national and International Conferences and in many peer-reviewed reputed journals. He has worked for many college committees. He has a good experience in academic coordination, conducting exam in college and training programs.

He is responsible for preparation of CS Program accreditation, as member of college Quality and Development Unit. Besides teaching, he enjoys giving tech talks and reading about new technology articles on internet. His teaching abilities include innovative skills and extensive use of technology in teaching.

Chapter 1.
The Internet of Things (IoT)

Building upon a complex network connecting billions of devices and humans into a multi-technology, multi-protocol and multi-platform infrastructure, the Internet-of-Things (IoT) main vision is to create an intelligent world where the physical, the digital and the virtual are converging to create smart environments that provide more intelligence to the energy, health, transport, cities, industry, buildings and many other areas of our daily life.

> A number of significant technology changes have come together to enable the rise of IoT. The prices of IoT hardware are dropping, putting sensors, processing power, network bandwidth, and cloud storage within reach of more users and making a wider range of IoT applications practical.

We define "the Internet of Things" as sensors and actuators connected by networks to computing systems. These systems can monitor or manage the health and actions of connected objects and machines. Connected sensors can also monitor the natural world, people, and animals.

The expectation is that of interconnecting millions of islands of smart networks enabling access to the information not only "anytime" and "anywhere" but also using "anything" and "anyone" ideally through any "path", "network" and "any service". This will be achieved by having the objects that we manipulate daily to be outfitted with sensing, identification and positioning devices and endowed with an IP address to become smart objects, capable of communicating with not only other smart objects but also with humans with the expectation of reaching areas that we could never reach without the advances made in the sensing, identification and positioning technologies.

We also observe the emergence of an Internet of Things ecosystem, another enabler of adoption. This includes vendors that specialize in IoT hardware and software, systems integrators, and a growing community of commercial and consumer IoT users.7 The actions of policy makers can advance or retard the evolution of the Internet of Things from this point. As we will discuss in Chapter 4, the potential economic impact that we estimate for IoT applications in 2025 depends on measures to make IoT data secure, protect personal privacy and intellectual property, and encourage interoperability among IoT devices and systems. Particularly in developing economies, low-cost data infrastructure is needed. Government agencies, working with technology providers, businesses, and consumers, can also participate in many of these efforts.

Finally, applying IoT technologies to human activities is already showing potential for massive change in people's lives. From giving people with chronic diseases new tools to manage their conditions to increasing fitness to avoid disease, the Internet of Things is beginning to demonstrate its potential to improve human health. Across the uses of IoT technology that we document in this report, people are the major beneficiaries—reducing their commuting times, offloading domestic chores to machines, saving money on energy, getting greater value from retail offers and in consumer products designed with IoT data, and enjoying life in safer homes and cities.

While being globally discoverable and queried, these smart objects can similarly discover and interact with external entities by querying humans, computers and other smart objects. The smart objects can also obtain intelligence by making or enabling context related decisions taking advantage of the available communication channels to provide information about themselves while also accessing information that has been aggregated by other smart objects.

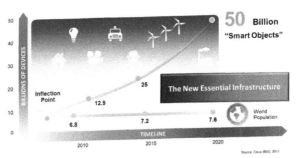

Figure 1. Internet-connected devices and the future evolution (Source: Cisco, 2011)

As revealed by Figure 1, the IoT is the new essential infrastructure which is predicted to connect 50 billion of smart objects in 2020 when the world population will reach 7.6 billion. As suggested by the ITU, such essential infrastructure will be built around a multi-layered architecture where the smart objects will be used to deliver different services through the four main layers depicted by Figure 2: a device layer, a network layer, a support layer and the application layer.

In the device layer lie devices (sensors, actuators, RFID devices) and gateways used to collect the sensor readings for further processing while the network layer provides the necessary transport and networking capabilities for routing the IoT data to processing places. The support layer is a middleware layer that serves to hide the complexity of the lower layers to the application layer and provide specific and generic services such as storage in different forms (database management systems and/or cloud computing systems) and many other services such as translation.

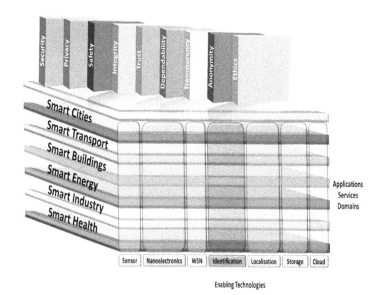

Figure 2. IoT Layered Architecture (Source: ITU-T)

The IoT can be perceived as an infrastructure driving a number of applications services which are enabled by a number of technologies. Its application services expand across many domains such as smart cities, smart transport, smart buildings, smart energy, smart industry and smart health while it is enabled by different technologies such as sensing, nanoeletronics, wireless sensor network (WSN), radio frequency identification (RFID), localization, storage and cloud. The IoT systems and applications are designed to provide security, privacy, safety, integrity, trust, dependability, transparency, anonymity and are bound by ethics constraints.

Experts say we are heading towards what can be called a "ubiquitous network society", one in which networks and networked devices are omnipresent. RFID and wireless sensors promise a world of networked and interconnected devices that provide relevant content and information whatever the location of the user. Everything from tires to toothbrushes will be in communications range, heralding the dawn of a new era, one in which today's Internet (of data and people) gives way to tomorrow's Internet of Things.

At the dawn of the Internet revolution, users were amazed at the possibility of contacting people and information across the world and across time zones. The next step in this technological revolution (connecting people any-time, anywhere) is to connect inanimate objects to a communication network. This vision underlying the Internet of things will allow the information to be accessed not only "anytime" and "anywhere" but also by "anything".

This will be facilitated by using WSNs and RFID tags to extend the communication and monitoring potential of the network of networks, as well as the introduction of computing power in everyday items such as razors, shoes and packaging.

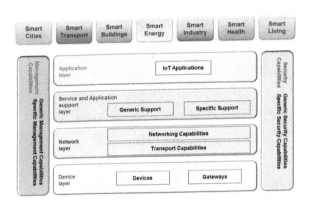

Figure 3 : IoT Application Stack

WSNs are an early form of ubiquitous information and communication networks. They are one of building blocks of the Internet of things.

Wireless Sensor Networks

A Wireless Sensor Network (WSN) is a self-configuring network of small sensor nodes (so-called motes) communicating among them using radio signals, and deployed in quantity to sense the physical world. Sensor nodes are essentially small computers with extremely basic functionality. They consist of a processing unit with limited computational power and limited memory, a radio communication device, a power source and one or more sensors.

Motes come in different sizes and shapes, depending on their foreseen use. They can be very small, if they are to be deployed in big numbers and need to have little visual impact. They can have a rechargeable battery power source if they are to be used in a lab. The integration of these tiny, ubiquitous electronic devices in the most diverse scenarios ensures a wide range of applications. Some of the application areas are environmental monitoring, agriculture, health and security.

In a typical application, a WSN is scattered in a region where it is meant to collect data through its sensor nodes. These networks provide a bridge between the physical world and the virtual world. They promise unprecedented abilities to observe and understand large scale, real-world phenomena at a fine spatio-temporal resolution. This is so because one deploys sensor nodes in large numbers directly in the field, where the experiments take place. All motes are composed of five main elements as shown below:

1. Processor: the task of this unit is to process locally sensed information and information sensed by other devices. At present the processors are limited in terms of computational power, but given Moore's law, future devices will come in smaller sizes, will be more powerful and consume less energy. The processor can run in different modes: sleep is used most of the time to save power, idle is used when data can arrive from other motes, and active is used when data is sensed or sent to / received from other motes.

2. Power source: motes are meant to be deployed in various environments, including remote and hostile regions so they must use little power. Sensor nodes typically have little energy storage, so networking protocols must emphasize power conservation. They also must have built-in mechanisms that allow the end user the option of prolonging network lifetime at the cost of lower throughput. Sensor nodes may be equipped with effective power scavenging methods, such as solar cells, so they may be left unattended for months, or years. Common sources of power are rechargeable batteries, solar panels and capacitors.

3. Memory: it is used to store both programs (instructions executed by the processor) and data (raw and processed sensor measurements).

4. Radio: WSN devices include a low-rate, short-range wireless radio. Typical rates are 10-100 kbps, and range is less than 100 meters. Radio communication is often the most power-intensive task, so it is a must to incorporate energy-efficient techniques such as wake-up modes. Sophisticated algorithms and protocols are employed to address the issues of lifetime maximization, robustness and fault tolerance.

5. Sensors: sensor networks may consist of many different types of sensors capable of monitoring a wide variety of ambient conditions. Table 1 classifies the three main categories of sensors based on field-readiness and scalability. While scalability reveals if the sensors are small and inexpensive enough to scale up to many distributed systems, the field-readiness describes the sensor's engineering efficiency with relation to field deployment. In terms of the engineering efficiency, Table 1 reveals high field-readiness for most physical sensors and for a few chemical sensors since most chemical sensors lie in the medium and low levels, while biological sensors have low field-readiness.

Sensor Category	Parameter	Field-Readiness	Scalability
Physical	Temperature	High	High
	Moisture Content	High	High
	Flow rate, Flow velocity	High	Med-High
	Pressure	High	High
	Light Transmission (Turb)	High	High
Chemical	Dissolved Oxygen	High	High
	Electrical Conductivity	High	High
	pH	High	High
	Oxydation Reduction Potential	Medium	High
	Major Ionic Species (Cl-, Na+)	Low-Medium	High
	Nutrientsa (Nitrate, Ammonium)	Low-Medium	Low-High
	Heavy metals	Low	Low
	Small Organic Compounds	Low	Low
	Large Organic Compounds	Low	Low
Biological	Microorganisms	Low	Low
	Biologically active contaminants	Low	Low

Common applications include the sensing of temperature, humidity, light, pressure, noise levels, acceleration, soil moisture, etc. Due to bandwidth and power constraints, devices primarily support low-data-units with limited computational power and limited rate of sensing.

Some applications require multi-mode sensing, so each device may have several sensors on board.

Following is a short description of the technical characteristics of WSNs that make this technology attractive.

1. **Wireless Networking**: motes communicate with each other via radio in order to exchange and process data collected by their sensing unit. In some cases, they can use other nodes as relays, in which case the network is said to be multi-hop. If nodes communicate only directly with each other or with the gateway, the network is said to be single-hop. Wireless

connectivity allows to retrieve data in real-time from locations that are difficult to access. It also makes the monitoring system less intrusive in places where wires would disturb the normal operation of the environment to monitor. It reduces the costs of installation: it has been estimated that wireless technology could eliminate up to 80 % of this cost.

2. **Self-organization**: motes organize themselves into an ad-hoc network, which means they do not need any pre-existing infrastructure. In WSNs, each mote is programmed to run a discovery of its neighborhood, to recognize which are the nodes that it can hear and talk to over its radio. The capacity of organizing spontaneously in a network makes them easy to deploy, expand and maintain, as well as resilient to the failure of individual points.

3. **Low-power**: WSNs can be installed in remote locations where power sources are not available. They must therefore rely on power given by batteries or obtained by energy harvesting techniques such as solar panels. In order to run for several months of years, motes must use low-power radios and processors and implement power efficient schemes. The processor must go to sleep mode as long as possible, and the Medium-Access layer must be designed accordingly. Thanks to these techniques, WSNs allow for long-lasting deployments in remote locations.

Applications of Wireless Sensor Networks

The integration of these tiny, ubiquitous electronic devices in the most diverse scenarios ensures a wide range of applications. Some of the most common application areas are environmental monitoring, agriculture, health and security. In a typical application, a WSN include:

1. Tracking the movement of animals. A large sensor network has been deployed to study the effect of micro climate factors in habitat selection of sea birds on Great Duck Island in Maine, USA. Researchers placed their sensors in burrows and used heat to detect the presence of nesting birds, providing invaluable data to biological researchers. The deployment was heterogeneous in that it employed burrow nodes and weather nodes.

2. Forest fire detection. Since sensor nodes can be strategically deployed in a forest, sensor nodes can relay the exact origin of the fire to the end users before the fire is spread uncontrollable. Researchers from the University of California, Berkeley, demonstrated the feasibility of sensor network technology in a fire environment with their FireBug application.

3. Flood detection. An example is the ALERT system deployed in the US. It uses sensors that detect rainfall, water level and weather conditions. These sensors supply information to a centralized database system.

4. Geophysical research. A group of researchers from Harvard deployed a sensor network on an active volcano in South America to monitor seismic activity and similar conditions related to volcanic eruptions.

5. Agricultural applications of WSN include precision agriculture and monitoring conditions that affect crops and livestock. Many of the problems in managing farms to maximize production while achieving environmental goals can only be solved with appropriate data. WSN can also be used in retail control, particularly in goods that require being maintained under controlled

conditions (temperature, humidity, light intensity, etc) [SusAgri].

6. An application of WSN in security is predictive maintenance. BP's Loch Rannoch project developed a commercial system to be used in refineries. This system monitors critical rotating machinery to evaluate operation conditions and report when wear and tear is detected. Thus one can understand how a machine is wearing and perform predictive maintenance. Sensor networks can be used to detect chemical agents in the air and water. They can also help to identify the type, concentration and location of pollutants.

7. An example of the use of WSN in health applications is the Bi-Fi, embedded system architecture for patient monitoring in hospitals and out-patient care. It has been conceived at UCLA and is based on the SunSPOT architecture by Sun. The motes measure high-rate biological data such as neural signals, pulse oximetry and electrocardiographs. The data is then interpreted, filtered, and transmitted by the motes to enable early warnings.

Roles in a Wireless Sensor Network
Nodes in a WSN can play different roles.

1. Sensor nodes are used to sense their surroundings and transmit the sensor readings to a sink node, also called "base station". They are typically equipped with different kinds of sensors. A mote is endowed with on-board processing, communication capabilities and sensing capabilities.

2. Sink nodes or "base stations" are tasked to collect the sensor readings of the other nodes and pass these readings to a gateway to which they are directly connected for further processing/analysis. A sink node is endowed with minimal on-board processing and communication capabilities but does not have sensing capabilities.

3. Actuators are devices which are used to control the environment, based on triggers revealed by the sensor readings or by other inputs. An actuator may have the same configuration as a mote but it is also endowed with controlling capabilities, for example to switch a light on under low luminosity.

Gateways often connected to sink nodes and are usually fed by a stable power supply since they consume considerable energy. These entities are normal computing devices such as laptops, notebooks, desktops, mobile phones or other emerging devices which are able to store, process and route the sensor readings to the processing place. However, they may not be endowed with sensing capabilities. Being range-limited, sensor motes require multi-hop communication capabilities to allow: 1) spanning distances much larger than the transmission range of a single node through localized communication between neighbor nodes 2) adaptation to network changes, for example, by routing around a failed node using a different path in order to improve performance and 3) using less transmitter power as a result of the shorter distance to be spanned by each node. They are deployed in three forms : (1) Sensor node used to sense the environment (2) Relay node used as relay for the sensor readings received from other nodes and (3) Sink node also often called base station which is connected to a gateway (laptop, tablet, iPod, Smart phone, desktop) with higher energy budget capable of either processing the sensor readings locally or to transmit these readings to remote processing places.

Chapter 2.
Internet Principles

Introduction to IPv6

IPv6 stands for Internet Protocol version 6, so the importance of IPv6 is implicit in its name, it's as important as the Internet! The Internet Protocol (IP from now on) was intended as a solution to the need to interconnect different data networks, and has become the "de facto" standard for all kinds of digital communications. Nowadays IP is present in most devices that are able to send and receive digital information, not only the Internet.

IP is standardized by the IETF (Internet Engineering Task Force), the organization in charge of all the Internet standards, guaranteeing the interoperability among software from different vendors. The fact that IP is a standard is of vital importance, because today everything is getting connected to the Internet using IP. All common Operating Systems and networking libraries support IP to send and receive data. As part of this "everything-connected-to-Internet" is the IoT, so now you know why you are reading this chapter about IPv6, the last version of the Internet Proto65437 In other words, today, the easiest way to send and receive data is by means of the standards used in the Internet, including IP.

A little bit of History

ARPAnet was the first attempt of the US Department of Defense (DoD) to devise a decentralized network more resilient to an attack, while able to interconnect completely different systems. ARPAnet was created in the seventies, but it was in 1983 when a brand new protocol stack was introduced, TCP/IP. The first widely used network protocol version was IPv4 (Internet Protocol version 4) which paved the way to the civilian Internet. Initially only research centers and universities were connected, supported by the NSF (National Science Foundation), and commercial applications where not allowed, but when the network started growing exponentially the NSF decided to transfer its operation and funding to private operators, lifting the restrictions to commercial traffic. While the main applications were email and file transfer, it was with the development of the World Wide Web based on the HTML protocol and specifically with the MOSAIC graphic interface browser and its successors that the traffic really exploded and the Internet began to be used by the masses. As a consequence there was a rapid depletion in the number of IP addresses available under IPv4, which was not designed to scale to these levels.

In order to allow for more addresses, you need a longer IP address space (greater number of bits to specify the address), which means a new architecture, which means changes to most of the routing and network software. After examining a number of proposals, the IETF settled on IPv6, described in the January 1995 RFC (Request for Comment, the official IETF documentation naming) 1752, sometimes also referred to as the Next Generation Internet Protocol, or IPng. The IETF updated the IPv6 standard in 1998 with the current definition covered in RFC 2460. By 2004, IPv6 was widely available from industry and supported by most new network equipment. Today IPv6 coexists with IPv4 in the Internet and the amount of IPv6 traffic is quickly growing as more and more ISPs and content providers have started supporting IPv6.

As you can see, the history of IP and Internet are almost the same, and because of this the growth of Internet is been hampered by the limitations of IPv4, and has led to the development of a new version of IP, IPv6, as the protocol to be used to interconnect all sorts of devices to send and/or receive information. There are even some technologies that are being developed only with IPv6 in mind, a good example in the context of the IoT is 6LowPAN.

From now on we will only center on IPv6. If you know something about IPv4, then you have half the way done, if not, don't worry we will cover the main concepts briefly and gently.

IPv6 Concepts

We will cover the the minimum you need to know about the last version of the Internet Protocol to understand why it's so useful for the IoT and how it's related with other protocols like 6LowPAN discussed later. We will assume that you are familiar with bits, bytes, networking stack, network layer, packets, IP header, etc. You should understand that IPv6 is a different protocol, non-compatible with IPv4.

In the following figure we represent the layered model used in the Internet.

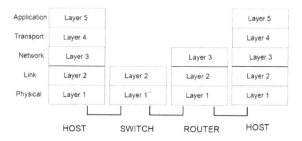

Figure 1.1. Internet Protocol stack

IPv6 sits in layer 3, called network layer. The pieces of data handled by layer 3 are called packets. Devices connected to the Internet can be hosts or routers. A host can be a PC, a laptop or a sensor board, sending and/or receiving data packets. Hosts will be the source or destination of the packets. Routers instead are in charge of packet forwarding, and are responsible of choosing the next router that will forward them towards the final destination. Internet is composed of a lot of interconnected routers, which receive data packets in one interface and send then as quick as possible using another interface towards another forwarding router.

IPv6 packet

The first thing you should know is what an IPv6 packet looks like. In the layered model we saw before, each layer introduces its own information in the packet, and this information is intended for, and can only be processed by the same layer in another IP device. This "conversation" between layers at the same level on different devices must follow a protocol.

The Internet layers are:

- **Application**: Here resides the software developed by programmers, that will use network services offered by the network stack. An example is the web browser that opens a network connection towards a web server. Another example is the web server software that runs in a server somewhere in the Internet waiting to answer request from client's browsers. Examples of application protocols are HTTP and DNS.
- **Transport**: Is a layer above the network layer that offers additional to it, for example, retransmission of lost packets or guaranteeing that the packets are received in the same order they were sent. This layer will be the one that shows a "network service" to the application layer, a service they can use to send or receive data. TCP and UDP are the most common transport protocols used in Internet.

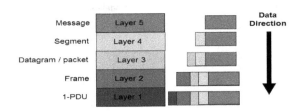

Figure 1.2. Data flow in the protocol stack

The bytes sent and received in the IP packet follow a standard format. The following figure shows the basic IPv6 header:

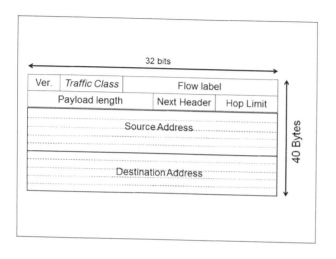

Figure 1.3. IPv6 Header

First you have the **basic IPv6 header** with a fixed size of 40 bytes, followed by upper layer data and optionally by some extension headers, which will be described later. As you can see there are several fields in the packet header, providing some improvements as compared with IPv4 header:

- The number of fields has been reduced from 12 to 8.
- The basic IPv6 header has a fixed size of 40 bytes and is aligned with 64 bits, allowing a faster hardware-based packet forwarding on routers.
- The size of addresses increased from 32 to 128 bits.

The most important fields are the source and destination addresses. As you already know, every IP device has a unique IP address that identifies it in the Internet. This IP address is used by routers to take their forwarding decisions.

IPv6 header has 128 bits for each IPv6 address, this allows for 2^{128} addresses (approximately 3.4×10^{38} ,i.e., 3.4 followed by 38 zeroes), whereas IPv4 uses 32 bits to encode each of the 2^{32} addresses (4,294,967,296) allowed.

We have seen the basic IPv6 header, and mentioned the **extension headers**. To keep the basic header simple and of a fixed size, additional features are added to IPv6 by means of extension headers.

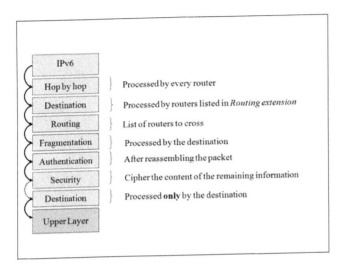

Figure 1.4. IPv6 Extension headers

Several extension headers have been defined, as you can see in the previous figure, and they have to follow the order shown. Extensions headers:

- Provide flexibility, for example, to enable security by ciphering the data in the packet.
- Optimize the processing of the packet, because with the exception of the hop by hop header, extensions are processed only by end nodes, (source and final destination of the packet), not by every router in the path.
- They are located as a "chain of headers" starting always in the basic IPv6 header, that use the field next header to point to the following extension header.

IPv6 addressing

1. The use of 128 bits for addresses brings some benefits:

- Provides many more addresses, to satisfy current and future needs, with ample space for innovation.
- Simplifies address auto-configuration mechanisms.
- Easier address management/delegation.

- Room for more levels of hierarchy and for route aggregation.

- Ability to do end-to-end IPsec.

IPv6 addresses are classified into the following categories (which also exist in IPv4): of four hexadecimal symbols, each group separated by a colon ":". The last two rules are for address notation compression, we will see how this works in the following.

Let's see some examples:

If we represent all the address bits we have the preferred form, for example:

2001:0db8:4004:0010:0000:0000:6543:0ffd

If we use squared brackets around the address we have the literal form of the address:

[2001:0db8:4004:0010:0000:0000:6543:0ffd]

If we apply the fourth rule, allowing compression within each group by eliminating leftmost zeroes, we have: 2001:db8:4004:10:0:0:6543:ffd

If we apply the fifth rule, allowing compression of one or more consecutive groups of zeroes using "::", we have: 2001:db8:4004:10::6543:ffd

Care should be taken when compressing and decompressing IPv6 addresses. The process should be reversible. It's very common to have some mistakes. For example, the following address 2001:db8:A:0:0:12:0:80 could be compressed even more using "::". we have two options:

a) 2001:db8:A::12:0:80 b) 2001:db8:A:0:0:12::80

Both are correct IPv6 addresses. But the address 2001:db8:A::12::80 is wrong, since it does not follow the last compression rule we saw above. The problem with this badly compressed address is that we can't be sure how to expand it, its ambiguous. We can't know if it expands to 2001:db8:A:0:12:0:0:80 or to 2001:db8:A:0:0:12:0:80 .

IPv6 network prefix

Last but not least you have to understand the concept of a network prefix, that indicates some fixed bits and some non-defined bits that could be used to create new sub-prefixes or to define complete IPv6 addresses assigned to hosts.
Let's see some examples:

1)The network prefix 2001:db8:1::/48 (the compressed form of 2001:0db8:0001:0000:0000:0000:0000:0000) indicates that the first 48 bits will always be the same (2001:0db8:0001) but that we can play with the other 80 bits, for example, to obtain two smaller prefixes: 2001:db8:1:a::/64 and 2001:db8:1:b::/64 .

2) If we take one of the smaller prefixes defined above, 2001:db8:1:b::/64 , where the first 64 bits are fixed we have the rightmost 64 bits to assign, for example, to an IPv6 interface in a host: 2001:db8:1:b:1:2:3:4 . This last example allow us to introduce a basic concept in IPv6:
* A /64 prefix is always used in a LAN (Local Area Network) .
*The rightmost 64 bits, are called the interface identifier (IID) because they uniquely identify

a host's interface in the local network defined by the /64 prefix. The following figure illustrates this statement:

Figure 1.6. Network and Interface ID

Now that you have seen your first IPv6 addresses we can enter into more detail about two types of addresses you will find when you start working with IPv6: reserved and unicast.

The unspecified address, used as a placeholder when no address is available:

0:0:0:0:0:0:0:0 (::/128)

The loopback address, is used by a node to send an IPv6 packet to itself:

0:0:0:0:0:0:0:1 (::1/128)

Documentation Prefix: 2001:db8::/32 . This prefix is reserved to be used in examples and documentation, you have already seen it in this chapter.

As specified in [RFC6890] IANA maintains a registry of special purpose IPv6 addresses [IANA-IPV6-SPEC].

The following are some other types of unicast addresses [RFC4291]:

- **Link-local**: Link-local addresses are always present in an IPv6 interface that is connected to a network. They all start with the prefix FE80::/10 and can be used to communicate with other hosts on the same local network, i.e., all hosts connected to the same switch. They cannot be used to communicate with other networks, i.e., to send or receive packets through a router.
- **ULA** (Unique Local Address) [RFC4193]: All ULA addresses start with the prefix FC00::/7, which in practice means that you could see FC00::/8 or FD00::/8 . Intended for local communications, usually inside a single site, they are not expected to be routable on the global Internet but used only inside a more limited environment.
- **Global Unicast**: Equivalent to the IPv4 public addresses, they are unique in the whole Internet and can be used to send a packet from one site to any destination in Internet.

What is IPv6 used for?

As we have seen IPv6 has some features that facilitates things like global addressing and host's address autoconfiguration. Because IPv6 provides as many addresses as we may need for some hundreds of years, we can put a global unicast IPv6 address on almost anything we may think of.

This brings back the initial Internet paradigm that every IP device could communicate with every IP device. This end-to-end communication allows bidirectional communication all over the Internet and between any IP device, which could result in collaborative applications and new ways of storing, sending and accessing the information.

In the context of this book we can, for example, contemplate IPv6 sensors all around the world collecting, sending and being accessed from different places to create a world-wide mesh of physical values measured, stored and processed.

The availability of a huge amount of addresses has allowed a new mechanism called **stateless address autoconfiguration** (SLAAC) that didn't exist with IPv4. Here is a brief summary of different ways to configure an address on an IPv6 interface:

- **Statically**: You can decide which address you will give to your IP device and then manually configure it into the device using any kind of interface: web, command line, etc. Normally you also have to configure other network parameters like the gateway to use to send packets out of your network.

- **DHCPv6** (Dynamic Host Configuration Protocol for IPv6) [RFC3315]: A porting of the similar mechanism already available in IPv4. You need to configure a dedicated server that after a brief negotiation with the device assigns an IP address to it. DHCPv6 allows IP devices to be configured automatically, this is why it is named as stateful address autoconfiguration, because the DHCPv6 server maintains a state of assigned addresses.

- **SLAAC**: Stateless address autoconfiguration [RFC4862] is a new mechanism introduced with IPv6 that allows to configure automatically all network parameters on an IP device using the router that gives connectivity to a network.

The advantage of SLAAC is that it simplifies the configuration of "dumb" devices, like sensors, cameras or any other device with low processing power. You don't need to use any interface in the IP device to configure anything, just "plug and net". It also simplifies the network infrastructure needed to build a basic IPv6 network, because you don't need additional device/server, you use the same router you need to send packets outside your network to configure the IP devices. We are not going to enter into details, but you just need to know that in a LAN (Local Area Network), connected to Internet by means of a router, this router is in charge of sending all the configuration information needed to its hosts using an RA (Router Advertisement) message. The router will send RAs periodically, but in order to expedite the process a host can send an RS (Router Solicitation) message when its interface gets connected to the network. The router will send an RA immediately in response to the RS.

The following figure show the packet exchange between a host that has just connected to a local network and some IPv6 destination in the Internet:

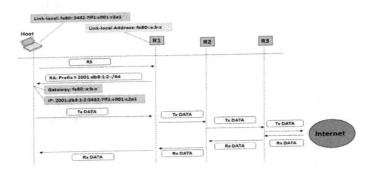

Figure 1.7. Packet exchange in IPv6

1) R1 is the router that gives connectivity to the host in the LAN and is periodically sending RAs.

2) Both R1 and Host have a link-local address in their interfaces connected to the host's LAN, this address is configured automatically when the interface is ready. Our host creates it's link-local address by combining the 64 leftmost bits of the link-local's prefix (fe80::/64) and the 64 rightmost bits of a locally generated IID (:3432:7ff1:c001:c2a1). These link-local addresses can be used in the LAN to exchange packets, but not to send packets outside the LAN.

3) The hosts needs two basic things to be able to send packets to other networks: a global IPv6 address and the address of a gateway, i.e., a router to which send the packets it wants to get routed outside its network.

4) Although R1 is sending RAs periodically (usually every several seconds) when the host get connected and has configured its link-local address, it sends an RS to which R1 responds immediately with an RA containing two things:

1. A **global prefix of length 64 bits** that is intended for SLAAC. The host takes the received prefix and adds to it a locally generated IID, usually the same one used for link-

local address. This way a global IPv6 address is configured in the host and now can communicate with the IPv6 Internet

2. Implicitly included is the **link-local address of R1**, because it is the source address of the RA. Our host can use this address to configure the **default gateway**, the place to which send the packets by default, to reach an IPv6 host somewhere in Internet.

5) Once both the gateway and global IPv6 address are configured, the host can receive or send information. In the figure it has something to send (Tx Data) to a host in Internet, so it creates an IPv6 packet with the destination address of the recipient host and as source a the just autoconfigured global address, which is sent to its gateway, R1's link-local address. The destination host can answer with some data (Rx Data).

Network Example

Following we show how a simple IPv6 network looks like, displaying IPv6 addresses for all the networking devices.

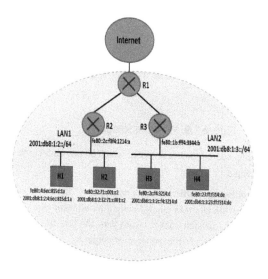

Figure 1.8. Simple IPv6 network

We have four hosts, (sensors, or other devices), and we want to put a pair of them in two different places, for example two floors in a building. We are dealing with four IP devices but you can have up to 2^{64} (18,446,744,073,709,551,616) devices connected on the same LAN.

We create two LANs with a router on each one,   bothrouters   connected to a central router (R1) that   provides connectivity to   Internet. LAN1 is   served by   R2 (with link-local address   fe80::2c:f3f4:1214:a   on that LAN) and uses the prefix 2001:db8:1:2::/64 announced by SLAAC. LAN2 is served by R3 (with link-local address fe80::1b:fff4:3344:b on that LAN) and uses the prefix 2001:db8:1:3::/64 announced by SLAAC.

All hosts have both a link-local IPv6 address and a global IPv6 address autoconfigured using the prefix provided by the corresponding router by means of RAs. In addition, remember that each host also configures the gateway using the link-local address used by the router for the RA. Link-local address can be used for communication among hosts inside a LAN, but for communicating with hosts in other LANs or any other network outside its own LAN a global IPv6 address is needed.

Short intro to Wireshark

Figure 1.9. Wireshark logo

Wireshark is a free and open-source packet analyzer, which allows packet traces to be sniffed, captured, and analyzed.

A packet trace is a record of traffic at some location on the network, as if a snapshot was taken of all the bits that passed across a particular wire. The packet trace records a timestamp for each packet, along with the bits that make up the packet, from the low-layer headers to the higher-layer contents.

Wireshark runs on most operating systems, including Windows, MAC and Linux. It provides a graphical user interface that shows the sequence of packets and the meaning of the bits when interpreted as protocol headers and data. The packets are color-coded to convey their meaning, and Wireshark includes various ways to filter and analyze them to let you investigate different aspects of behavior. It is widely used to troubleshoot networks.

A common usage scenario is when a person wants to troubleshoot network problems or look at the internal workings of a network protocol. A user could, for example, see exactly what happens when he or she opens up a website or sets up a wireless sensor network. It is also possible to filter and search for given packet attributes, which facilitates the debugging process.

More information and installation instructions are available at Wireshark site..

Figure 1.10. Wireshark Screenshot

When you open Wireshark, there are four main areas, from top to bottom: menus and filters, list of captured packets, detailed information about the selected packet, including its full content in hexadecimal and ASCII. Online directly links you to the Wiresharks site, where you can find a handy user guide and information on the security of Wireshark. Under Files, you'll find Open, which lets you open previously captured files,, and Sample Captures. You can download any of the sample captures through this website, and study the data. This will help you understand what kind of packets Wireshark can capture.

The Capture section let you choose your Interface from the available ones. It'll also show you which ones are active. Clicking details will show you some pretty generic information about that interface.

Under Start, you can choose one or more interfaces to check out. Capture Options allows you to customize what information you see during a capture. Here you can choose a filter, a capture file, and more. Under Capture Help, you can read up on how to capture, and you can check info on Network Media about which interfaces work on which platforms.

Let's select an interface and click Start. To stop a capture, press the red square in the top toolbar. If you want to start a new capture, hit the green triangle which looks like a shark fin next to it. Now that you have got a finished capture, you can click File, and save, open, or merge the capture. You can print it, you can quit the program, and you can export your packet capture in a variety of ways.

You can find a certain packet, copy packets, mark (highlight) any specific packet or all the packets. Another interesting thing you can do under Edit, is resetting the time value. You'll notice that the time is in seconds incrementing. You can reset it from the packet you've clicked on. You can add a comment to a packet, configure profiles and preferences.

When we select a packet from the list of captured ones, Wireshark shows detailed information of the different protocols used by that packet, for example Ethernet:

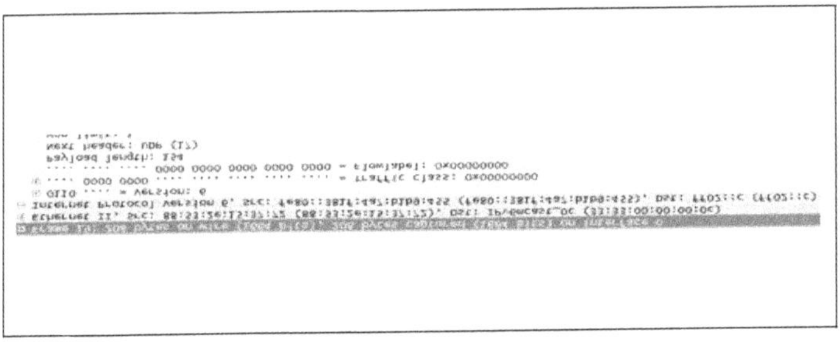

Figure 1.11. Ethernet packet

Or IPv6, where we can see the fields we mentioned before: Version, Traffic class, flowlabel, payload length, next header, etc.:

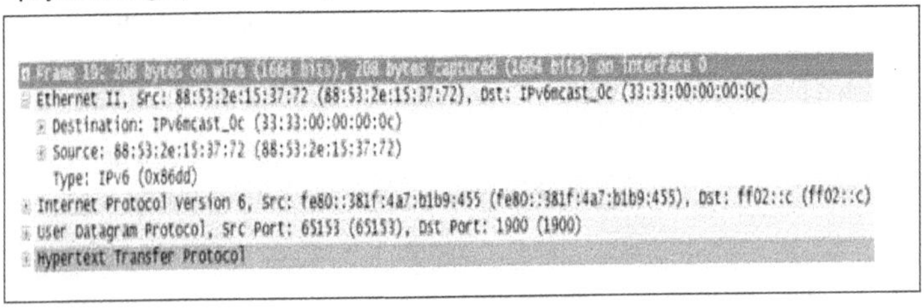

Figure 1.12. IPv6 packet

There are two methods to apply filters to the list of captured packets:

- Write a filter expression on the specific box and then apply it. Protocols can be specified (ip,ipv6, icmp, icmpv6), fields of a protocol (ipv6.dst, ipv6.src) and even complex expressions can be created using operators like AND (&&), OR (||) or the negation (|).

- Another option to create filters is to right click in one filed of a captured packet, in the list of captured packets. There will appear a menu option "Apply as filter", with several options on how to use that field.

Figure 1.14. Wireshark Captured packets

Another useful and interesting option of Wireshark is the possibility to see statistics about the captured traffic. If we have applied filters, the statistics will be about the filtered traffic. Just go to the Statistics menu and select, for example, Protocol Hierarchy:

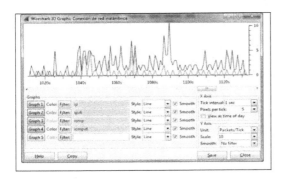

Figure 1.15. Wireshark statististic

Other interesting options are:

- Conversation List → IPv6

- Statistics → Endpoint List → IPv6

- Statistics → IO Graph

This last option allow to create graphs with different lines for different types of traffic and save the image:

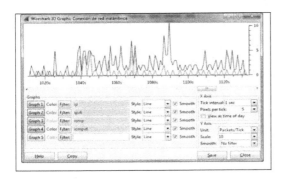

Figure 1.16. Wireshark charts

IPv6 Exercises

Let's test your IPv6 knowledge with the following exercises:
1) What is the size of IPv4 and IPv6 addresses?
   a. 32-bits, 128-bits
   b. 32-bits, 64-bits
   c. 32-bits, 112-bits
   d. 32-bits, 96-bits
   e. none of these

2) Which of the following is a valid IPv6 address notation rule?
   a. Zeroes on the right inside a group of 16 bits can be eliminated
   b. The address is divided in 5 groups of 16 bits separated by ":"
   c. The address is divided in 8 groups of 16 bits separated by "."
   d. One or more groups of all zeroes could be substituted by "::"
   e. Decimal notation is used grouping bits in 4 (nibbles)

3) Interface Identifiers (IID) or the rightmost bits of an IPv6 address used on a LAN will be 64 bits long.
   a. True
   b. False

4) Which of the following is a correct IPv6 address?
   a. 2001:db8:A:B:C:D::1
   b. 2001:db8:000A:B00::1:3:2:F
   c. 2001:db8:G1A:A:FF3E::D
   d. 2001:0db8::F:A::B

5) Which ones of the following sub-prefixes belong to the prefix 2001:db8:0A00::/48 ? (Choose all that apply)
   a. 2001:db9:0A00:0200::/56
   b. 2001:db8:0A00:A10::/64
   c. 2001:db8:0A:F:E::/64
   d. 2001:db8:0A00::/64

6) IPv6 has a basic header with more fields than IPv4 header
   a. True
   b. False

7) Extension headers can be added in any order
   a. True
   b. False

8) Autoconfiguration of IP devices is the same in IPv4 and IPv6
   a. True
   b. False

9) Which one is not an option for configuring an IPv6 address in an interface?
   a. DHCPv6

b. Fixed address configured by vendor
c. Manually
d. SLAAC (Stateless Address Autoconfiguration)

10) Which packets are used by SLAAC to autoconfigure an IPv6 host?
a. NS/NA (Neighbor Solicitation / Neighbor Advertisement)
b. RS/RA (Router Solicitation / Router Advertisement)
c. Redirect messages
d. NS / RA (Neighbor Solicitation / Router Advertisement)

Addressing Exercises

A) Use the two compression rules for the utmost compression of the following addresses:
1. 2001:0db8:00A0:7200:0fe0:000B:0000:0005
2. 2001:0db8::DEFE:0000:C000
3. 2001:db8:DAC0:0FED:0000:0000:0B00:12

B) Apply maximum decompression (representing all the 32 nibbles in hexadecimal) to the following addresses:
1. 2001:db8:0:50::A:123
2. 2001:db8:5::1
3. 2001:db8:C00::222:0CC0

C) You receive the following IPv6 prefix for your network: 2001:db8:A:0100::/56 , shown in the following figure:

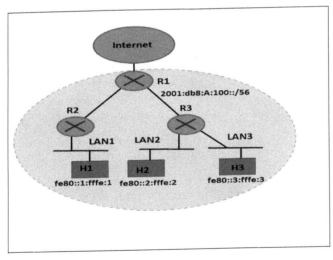

Figure 1.17. LAN Example

Please determine:

a. IPv6 prefix for LAN1, a /64 prefix taken from the /56 you have.
b. IPv6 prefix for LAN2, a /64 prefix taken from the /56 you have.
c. IPv6 prefix for LAN3, a /64 prefix taken from the /56 you have.
d. A global IPv6 address using the LAN1 prefix for H1 host (added to the link-local address already used).
e. A global IPv6 address using the LAN2 prefix for H2 host (added to the link-local address already used).
f. A global IPv6 address using the LAN3 prefix for H3 host (added to the link-local address already used).

> Hint: To divide the /56 prefix into /64 prefixes, you have to change the value of the bits 57 to 64, i.e., the XY values in 2001:db8:A:01XY::/64 .

Connecting our IPv6 Network to the Internet

As said in the introduction of this book, network communications is one of the four basic elements of an IoT system. We already have seen that IPv6 brings the possibility of giving an IP address to almost anything we can think of, and can do this making it easy to autoconfigure network parameters on our devices.

Once we have all our "things" connected using IPv6, they can use it to communicate among them locally or with any other "thing" on the IPv6 Internet. In this chapter we will **focus on the Internet side of the communication of the "things"** composing the Internet of Things.

As we will see in this book, the capability of connecting our devices to the Internet allows new possibilities and services. For example, we can connect our wireless sensors networks to a centralized repository, where all the sensed information can be processed and stored for historical records, which will uncover underlying patterns and maybe predict future events. This basic idea is what nowadays is called "Big Data" and has a whole set of its own concepts and techniques.

Wireless IoT Communication Protocols

IoT involves smart devices available in the entire region and networked them as personal wireless network locally and worldwide through the Internet. wireless communication technologies required for connectivity. Some main protocols are Bluetooth, Wi-Fi, ZigBee, Z-Wave and RF Link. In IOT implementation Communication Technologies utilized by IoT devices are summarized below:

i. ZigBee

This ZigBee standard defines the physical and Medium Access Control layers for cheap wireless networks [3]. The physical layer Zigbee functions are channel selection, link quality, energy measurement and channel assessment. For the network and the application layer, ZigBee standard is applicable. The network layer provides routing over the internet, specifying different network topologies: star, tree, peer-to-peer and mesh. In the application layer provides a framework for distributed application development and communication. Zigbee is used in agriculture and food industry, additionally used in a smart home, automation, security, and medical monitoring. [4]

ii. RF Links

RF communication modules data rates are quite low ranges up to 1Mpbs and also need an Internet-enabled gateway that will provide access to the devices for making a complete IoT network. The Radio Frequency Identification (RFID) technology has been initially introduced for identifying and tracking objects with the help of small electronic chips, called tags. RFID has been originally categorized as the enabling communication power for the Internet of Things, due to its low cost, high mobility, and efficiency in identifying devices and objects. Despite RFID is very common for device identification and some information exchange [5]

iii. Bluetooth

Bluetooth (IEEE 802.15.1) is a wireless protocol for short-range communication in wireless personal area networks (PAN) as a cable replacement for mobile devices. It uses the 2.4 GHz radio bands to communicate at 1 Mb per second between up to seven devices. Bluetooth is low-powered devices to use for small data like health or tracker. When connectivity is initiated Bluetooth comes into action from sleep mode and conserves power. It uses a method of frequency-hopping spread-spectrum (FHSS) communication, transmits data over different frequencies at different time intervals. Bluetooth uses a master-slave-based MAC (medium access control) protocol [6-8]

iv. 6LoWPAN

The 6LoWPAN is Wireless PAN with low power, large-scale network and supports IPv6. It is a connection-oriented technology in which router forwards the data to its next hop to the 6LoWPAN gateway which is connected to 6LoWPAN with the IPv6 domain and then forwards the data to its respected device correctly. With IPv6 we have enough address space to identify all the things in the world. In IP based network standard protocols (HTTP, TCP/IP) are directly applied to sensor nodes just as they do with traditional web servers out there on the Internet [9][10].

v. Z-Wave

Z-Wave protocol architecture is low power consuming mostly used in home automation, security and light commercial environment. It has an open communication protocol. The main purpose of Z-wave is a reliable mesh network, message passing from a control unit to one or more nodes in the network. Z-wave have two types of devices, one is poll Controllers which send commands to the slaves, the second type of device, which replies to the controller to execute the commands.

vi. Wi-Fi Wireless fidelity

Wireless fidelity is known as Wi-Fi, with the IEEE 802.11x standards, common way to connect devices wirelessly to the Internet. Laptop, Smartphone, and Tablet PC are equipped with Wi-Fi interfaces and talk to a wireless router and provide two-way accesses to the Internet. The Wi-Fi standard family allows establishing a wireless network on short distances. Wi-Fi has series types of networks of IEEE 802.11 and IEEE security extension.

The Wi-Fi group is working on an unlicensed spectrum of 2.4 GHz band. WIFI consumes more power.

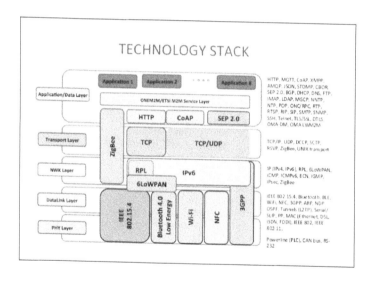

Figure: IOT research EU

MQTT - Message- Queue Telemetry- Transport

It is a publish subscribe based lightweight messaging protocol for use in conjunction with the TCP/IP protocol. This provides an embedded connection from applications to a middleware's on one side. on the other
side connection from networks to communications, The system consists of three main components: publishers, subscribers, and an agent. MQTT enters in sleep mode once data transmission complete.

CoAP - Constrained Application Protocol

CoAP designed for small devices, Machine to Machine, low powered applications such as smart energy and building automation.Request-Response based model for end-points using Client-Server interaction is asynchronous over a datagram-oriented transport protocol such as UDP.

XMPP – Extensible Messaging and Presence Protocol

Extensible Messaging and Presence Protocol (XMPP) is a messaging protocol is a highly efficient protocol over the internet. XMPP supports both publish/ subscribe and request/ response architecture and it is up to the application developer to choose which architecture to use. Real-time application and M2M can be implemented using XMPP having low latency.

Data Handling in IOT

Data management in IOT is at various levels is handled by

- Sensor
- Sensor cloud (FoG)
- Cloud

Cloud model cannot store all these IOT data. Data need to be filtered. The ability of the current cloud model is insufficient to handle the requirements of IoT. Issues are: Volume, Latency, and Bandwidth

Volume: The total number of connected vehicles worldwide will be 250 million by 2020. There will be more than 30 billion IoT devices. The amount of data generated by IoT devices is simply huge. Private firms, Factories, airplane companies' produce colossus amount of data every day.

Latency: Latency is time taken by a data packet of a round trip. Latency is an important aspect of handing a time-sensitive data. If edge devices send time-sensitive data to the cloud for analysis and wait for the cloud to give a proper action, then it can lead to many unwanted results. While handling time-sensitive data, a millisecond can make huge differences. Reduce latency of data and appropriate actions at the right time prevent major accidents machine failure etc. A minute delay while taking a decision makes a huge difference. Latency can be reduced by analyzing the data close to the data source

Bandwidth: Bandwidth is bit-rate of data during transmission. If all the data generated by IoT devices are sent to cloud for storage and analysis, then, the traffic generated by these devices will be simply enormous, consumes almost all the bandwidths. Handling this kind of traffic will be simply a very hard task.

Getting back to the network connectivity domain, our objective is to connect IoT devices to the Internet using IPv6, allowing communication with other IoT devices, collecting servers or even with people.

Related with the IPv6 connectivity to Internet is an important idea: **communication between IoT devices and the IPv6 Internet could be bidirectional**. This is important to remark because with IPv4, connectivity is oftentimes designed as a one direction channel between a client and a server. This changes with IPv6.

Having a bidirectional communication with the IoT devices allows useful possibilities, because its not just that the device can send information to somewhere in the Internet, but that anybody in the Internet could be able to send information, requests or commands to the IoT device. This can be used in different scenarios:

- **Management**: To manage the IoT device performing some status tests, updating some parameters/configuration/firmware remotely allowing for a better and efficient use of the hardware platform and improving the infrastructure security.

- **Control**: Send commands or control actuators to make the IoT device perform an action.

- **Communication**: Send information to the IoT device, that can be displayed using some kind of interface.

IIPv6 is still being deployed all over the different networks that compose the Internet, which means that different scenarios can be found when deciding how to connect our network to the IPv6 Internet. Following are the three most common scenarios, in preferred order, being Native IPv6 connectivity the best choice.

- **Native IPv6 Connectivity**: This scenario applies when both the ISP providing connectivity to the Internet and the router(s) and networks devices used in our network support of IPv6. Native IPv6 means that the IPv6 packets will flow without being changed or tunnelled anywhere in its path from origin to destination. It is common to find what is called dual-stack networks, where both native IPv6 and native IPv4 are being used at the same time in the same interfaces and devices. This native IPv6 scenario covers both cases: IPv6-only and dual-stack.

- **No IPv6 connectivity**: In this scenario we face a common problem nowadays, the lack of IPv6 connectivity from an ISP. Although we have IPv6 support on the router that connects our network to Internet, the ISP supports only IPv4. The solution is to use one of the so called IPv6 Transition Mechanism. The most simple and useful in this case would be the 6in4 tunnel, based on creating a point-to-point static tunnel that encapsulates IPv6 packets into IPv4.

- **No IPv6 connectivity and no IPv6 capable router**: This scenario covers the case where there is no IPv6 connectivity from the ISP, nor IPv6 support on the router connecting our network to the Internet. As seen before, to solve the lack of IPv6 connectivity from the ISP we can use a 6in4 tunnel, but in this scenario we also have to face the lack of IPv6 support on the router which prevents the creation of the tunnel. The solution is to add a new router that supports both IPv6 and IPv4, and create a 6in4 tunnel from this router to a tunnel end router somewhere on the IPv4 Internet.

-

Chapter 3.
Introduction to 6LoWPAN

One of the drivers of the IoT, where anything can be connected, is the use of wireless technologies to create a communication channel to send and receive information. This wide adoption of wireless technologies allows increasing the number of connected devices but results in limitations in terms of cost, battery life, power consumption, and communication distance for the devices. New technologies and protocols should tackle a new environment, usually called Low power and Lossy networks (LLNs), with the following characteristics:

1. Significantly more devices than those on current local area networks.

2. Severely limited code and ram space in devices.

3. Networks with limited communications distance (range), power and processing resources.

4. All elements should work together to optimize energy consumption and bandwidth usage.

Another factor that is being widely adopted within IoT is the use of IP as the network protocol. The use of IP provides several advantages, because it is an open standard that is widely available, allowing for easy and cheap adoption, good interoperability and easy application layer development. The use of a common standard like an end-to-end IP-based solution avoids the problem of non-interoperable networks.

For wireless communication technology, the IEEE 802.15.4 standard [IEEE802.15.4] is very promising for the lower (link and physical) layers, although others are also being considered as good options like Low Power WiFi, Bluetooth ® Low Energy, DECT Ultra Low Energy, ITU-T G.9959 networks, and NFC (Near Field Communication).

One component of the IoT that has received significant support from vendors and standardization organizations is that of WSN (Wireless Sensor Networks).

The IETF has different working groups (WGs) developing standards to be used by WSN:

1. **6lowpan**: IPv6 over Low-power Wireless Personal Area Networks [sixlowpan], defines the standards for IPv6 communication over the IEEE 802.15.4 wireless communication technology. 6lowpan acts as an adaptation layer between the standard IPv6 world and the low power and lossy wireless communications medium offered by IEEE 802.15.4. Note that this standard is only defined with IPv6 in mind, no IPv4 support is available.

2. **roll**: Routing Over Low power and Lossy networks [roll]. LLNs have specific routing requirements that could not be satisfied with existing routing protocols. This WG focuses on routing solutions for a subset of all possible application areas of LLNs (industrial, connected home, building and urban sensor networks), and protocols are designed to satisfy their

application-specific routing requirements. Here again the WG focuses only on the IPv6 routing architectural framework.

3. **6lo**: IPv6 over Networks of Resource-constrained Nodes [sixlo]. This WG deals with IPv6 connectivity over constrained node networks. It extends the work of the 6lowpan WG, defining IPv6-over-foo adaptation layer specifications using 6LoWPAN for link layer in constrained node networks.

As seen, 6LoWPAN is the basis of the work carried out in standardization at IETF to communicate constrained resources nodes in LLNs using IPv6. The work on 6LoWPAN has been completed and is being further complemented by the roll WG to satisfy routing needs and the 6lo WG to extend the 6lowpan standards to any other link layer technology. Following are more details about 6LoWPAN, as the first step into the IPv6 based WSN/IoT. 6LoWPAN and related standards are concerned about providing IP connectivity to devices, irrelevantly of the upper layers, except for the UDP transport layer protocol that is specifically considered.

Overview of LoWPANs

Low-power and lossy networks (LLNs) is the term commonly used to refer to networks made of highly constrained nodes (limited CPU, memory, power) interconnected by a variety of "lossy" links (low-power radio links). They are characterized by low speed, low performance, low cost, and unstable connectivity.

A LoWPAN is a particular instance of an LLN, formed by devices complying with the IEEE 802.15.4 standard.

The typical characteristics of devices in a LoWPAN are:

1. **Limited Processing Capability**: Different types and clock speeds processors, starting at 8-bits.
2. **Small Memory Capacity**: From few kilobytes of RAM with a few dozen kilobytes of ROM/ flash memory, it's expected to grow in the future, but always trying to keep at the minimum necessary.
3. **Low Power**: In the order of tens of milliamperes.
4. **Short Range**: The Personal Operating Space (POS) defined by IEEE 802.15.4 implies a range of 10 meters. For real implementations it can reach over 100 meters in line-of-sight situations.
5. **Low Cost**: This drives some of the other characteristics such as low processing, low memory, etc.

All this constraints on the nodes are expected to change as technology evolves, but compared to other fields it's expected that the LoWPANs will always try to use very restricted devices to allow for low prices and long life which implies hard restrictions in all other features.

A LoWPAN typically includes devices that work together to connect the physical environment to real-world applications, e.g., wireless sensors, although a LoWPAN is not necessarily comprised of sensor nodes only, since it may also contain actuators.

It's also important to identify the characteristics of LoWPANs, because they will be the constraints guiding all the technical work:

1. Small packet size: Given that the maximum physical layer frame is 127 bytes, the resulting maximum frame size at the media access control layer is 102 octets. Link-layer security imposes further overhead, which leaves a maximum of 81 octets for data packets.
2. IEEE 802.15.4 defines several addressing modes: It allows the use of either IEEE 64-bit extended addresses or (after an association event) 16-bit addresses unique within the PAN (Personal Area Network).
3. Low bandwidth: Data rates of 250 kbps, 40 kbps, and 20 kbps for each of the currently defined physical layers (2.4GHz, 915MHz, and 868MHz, respectively).
4. Topologies include star and mesh.
5. Large number of devices expected to be deployed during the lifetime of the technology. Location of the devices is typically not predefined, as they tend to be deployed in an ad-hoc fashion. Sometimes the location of these devices may not be easily accessible or they may move to new locations.
a. The pervasive nature of IP networks allows leveraging existing infrastructure.
b. IP-based technologies already exist, are well-known, proven to be working and widely available. This allows for an easier and cheaper adoption, good interoperability and easier application layer development.
c. IP networking technology is specified in open and freely available specifications, which is able to be better understood by a wider audience than proprietary solutions.
d. Tools for IP networks already exist.
e. IP-based devices can be connected readily to other IP-based networks, without the need for intermediate entities like protocol translation gateways or proxies.
f. The use of IPv6, specifically, allows for a huge amount of addresses and provides for easy network parameters autoconfiguration (SLAAC). This is paramount for 6LoWPANs where large number of devices should be supported.

On the counter side using IP communication in LoWPANs raise some issues that should be taken into account:

a. IP Connectivity: One of the characteristics of 6LoWPANs is the limited packet size, which implies that headers for IPv6 and layers above must be compressed whenever possible.
b. Topologies: LoWPANs must support various topologies including mesh and star: Mesh

topologies imply multi-hop routing to a desired destination. In this case, intermediate devices act as packet forwarders at the link layer. Star topologies include provisioning a subset of devices with packet forwarding functionality. If, in addition to IEEE 802.15.4, these devices use other kinds of network interfaces such as Ethernet or IEEE 802.11, the goal is to seamlessly integrate the networks built over those different technologies. This, of course, is a primary motivation to use IP to begin with.

c. Limited Packet Size: Applications within LoWPANs are expected to originate small packets. Adding all layers for IP connectivity should still allow transmission in one frame, without incurring excessive fragmentation and reassembly. Furthermore, protocols must be designed or chosen so that the individual "control/protocol packets" fit within a single 802.15.4 frame.

d. Limited Configuration and Management: Devices within LoWPANs are expected to be deployed in exceedingly large numbers. Additionally, they are expected to have limited display and input capabilities. Furthermore, the location of some of these devices may be hard to reach. Accordingly, protocols used in LoWPANs should have minimal configuration, preferably work "out of the box", be easy to bootstrap, and enable the network to self heal given the inherent unreliable characteristic of these devices.

e. Service Discovery: LoWPANs require simple service discovery network protocols to discover, control and maintain services provided by devices.

f. Security: IEEE 802.15.4 mandates link-layer security based on AES, but it omits any details about topics like bootstrapping, key management, and security at higher layers. Of course, a complete security solution for LoWPAN devices must consider application needs very carefully.

6LoWPAN

We have seen that there is a lower layer (physical and link layers on TCP/IP stack model) that provide connectivity to devices in what is called a LoWPAN. Also that using IPv6 over this layer would bring several benefits. The main reason for developing the IETF standards mentioned in the introduction is that between the IP (network layer) and the lower layer there are some important issues that need to be solved by means of an adaptation layer, the 6lowpan.

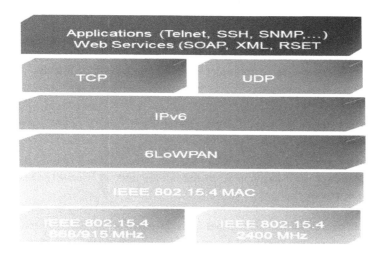

Figure 2.1. 6LoWPAN in the protocol stack

The main goals of 6lowpan are:

1. Fragmentation and Reassembly layer: IPv6 specification [RFC2460] establishes that the minimum MTU that a link layer should offer to the IPv6 layer is 1280 bytes. The protocol data units may be as small as 81 bytes in IEEE 802.15.4. To solve this difference a fragmentation and reassembly adaptation layer must be provided at the layer below IP.

2. Header Compression: Given that in the worst case the maximum size available for transmitting IP packets over an IEEE 802.15.4 frame is 81 octets, and that the IPv6 header is 40 octets long, (without optional extension headers), this leaves only 41 octets for upper-layer protocols, like UDP and TCP. UDP uses 8 octets in the header and TCP

uses 20 octets. This leaves 33 octets for data over UDP and 21 octets for data over TCP. Additionally, as pointed above, there is also a need for a fragmentation and reassembly layer, which will use even more octets leaving very few octets for data. Thus, if one were to use the protocols as is, it would lead to excessive fragmentation and reassembly, even when data packets are just 10s of octets long. This points to the need for header compression.

3. Address Autoconfiguration: specifies methods for creating IPv6 stateless address auto configuration (in contrast to stateful) that is attractive for 6LoWPANs, because it reduces the configuration overhead on the hosts. There is a need for a method to generate the IPv6 IID (Interface Identifier) from the EUI-64 assigned to the IEEE 802.15.4 device.

4. Mesh Routing Protocol: A routing protocol to support a multi-hop mesh network is necessary. Care should be taken when using existing routing protocols (or designing new ones) so that the routing packets fit within a single IEEE 802.15.4 frame. The mechanisms defined by 6lowpan IETF WG are based on some requirements for the IEEE 802.15.4 layer:

5. IEEE 802.15.4 defines four types of frames: beacon frames, MAC command frames, acknowledgement frames and data frames. IPv6 packets must be carried on data frames.

6. Data frames may optionally request that they be acknowledged. It is recommended that IPv6 packets be carried in frames for which acknowledgements are requested so as to aid link-layer recovery.

7. The specification allows for frames in which either the source or destination addresses (or both) are elided. Both source and destination addresses are required to be included in the IEEE 802.15.4 frame header.

8. The source or destination PAN ID fields may also be included. 6LoWPAN standard assumes that a PAN maps to a specific IPv6 link.

9. Both 64-bit extended addresses and 16-bit short addresses are supported, although additional constraints are imposed on the format of the 16-bit short addresses.

10. Multicast is not supported natively in IEEE 802.15.4. Hence, IPv6 level multicast packets must be carried as link-layer broadcast frames in IEEE 802.15.4 networks. This must be done such that the broadcast frames are only heeded by devices within the specific PAN of the link in question.

The 6LoWPAN adaptation format was specified to carry IPv6 datagrams over constrained links, taking into account limited bandwidth, memory, or energy resources that are expected in applications such as wireless sensor networks. For each of these goals and requirements there is a solution provided by the 6lowpan specification:

1. A Mesh Addressing header to support sub-IP forwarding.

2. A Fragmentation header to support the IPv6 minimum MTU requirement.

3. A Broadcast Header to be used when IPv6 multicast packets must be sent over the IEEE 802.15.4 network.

4. Stateless header compression for IPv6 datagrams to reduce the relatively large IPv6 and UDP headers down to (in the best case) several bytes. These header are used as the LoWPAN encapsulation, and could be used at the same time forming what is called the header stack. Each header in the header stack contains a header type followed by zero or more header fields. When more than one LoWPAN header is used in the same packet, they must appear in the following order: Mesh Addressing Header, Broadcast Header, and Fragmentation Header.

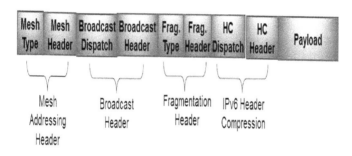

Figure 2.2. 6LoWPAN headers

IPv6 Interface Identifier (IID)

As already said an IEEE 802.15.4 device could have two types of addresses. For each one there is a different way of generating the IPv6 IID.

1. IEEE EUI-64 address: All devices have this one. In this case, the Interface Identifier is formed from the EUI-64, complementing the "Universal/Local" (U/L) bit, which is the next-to-lowest order bit of the first octet of the EUI-64. Complementing this bit will generally change a 0 value to a 1.

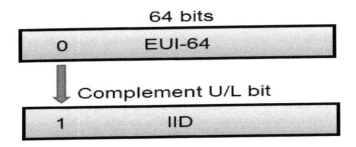

Figure 2.3. EUI-64 derived IID

2. 16-bit short addresses: Possible but not always used. The IPv6 IID is formed using the PAN (or zeroes in case of not knowing the PAN) and the 16 bit short address as in the figure below.

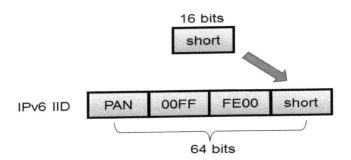

Figure 2.4. IPv6IID

Header Compression

Two encoding formats are defined for compression of IPv6 packets: LOWPAN_IPHC and LOWPAN_NHC, an encoding format for arbitrary next headers.

To enable effective compression, LOWPAN_IPHC relies on information pertaining to the entire 6LoWPAN. LOWPAN_IPHC assumes the following will be the common case for 6LoWPAN communication:

1. Version is 6.

2. Traffic Class and Flow Label are both zero.

3. Payload Length can be inferred from lower layers from either the 6LoWPAN Fragmentation header or the IEEE 802.15.4 header.

4. Hop Limit will be set to a well-known value by the source.

5. Addresses assigned to 6LoWPAN interfaces will be formed using the link-local prefix or a small set of routable prefixes assigned to the entire 6LoWPAN.

6. Addresses assigned to 6LoWPAN interfaces are formed with an IID derived directly from either the 64-bit extended or the 16-bit short IEEE 802.15.4 addresses. Depending on how closely the packet matches this common case, different fields may not be compressible thus needing to be carried "in-line" as well. The base format used in LOWPAN_IPHC encoding is shown in the figure below.

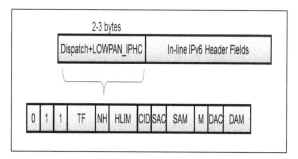

Figure 2.5. Header compression

Where:

- TF: Traffic Class, Flow Label.

- NH: Next Header.

- HLIM: Hop Limit.

- CID: Context Identifier Extension.

- SAC: Source Address Compression.

- SAM: Source Address Mode.

- M: Multicast Compression.
- DAC: Destination Address Compression.

- DAM: Destination Address Mode.

Not going into details, it's important to understand how 6LoWPAN compression works. To this end, let's see two examples:

1. HLIM (Hop Limit): Is a two bits field that can have four values, three of them make the hop limit field to be compressed from 8 to 2 bits:

   a. 00: Hop Limit field carried in-line. There is no compression and the whole field is carried in-line after the LOWPAN_IPHC.

   b. 01: Hop Limit field compressed and the hop limit is 1.

   c. 10: Hop Limit field compressed and the hop limit is 64.

   d. 11: Hop Limit field compressed and the hop limit is 255.

2. SAC/DAC used for the source IPv6 address compression. SAC indicates which address compression is used, stateless (SAC=0) or stateful context-based (SAC=1). Depending on SAC, DAC is used in the following way:

   a. If SAC=0, then SAM:

      - 00: 128 bits. Full address is carried in-line. No compression.

      - 01: 64 bits. First 64-bits of the address are elided, the link-local prefix. The remaining 64 bits are carried in-line.

      - 10: 16 bits. First 112 bits of the address are elided. First 64 bits is the link-local prefix. The following 64 bits are 0000:00ff:fe00:XXXX, where XXXX are the 16 bits carried in-line.

      - 11: 0 bits. Address is fully elided. First 64 bits of the address are the link-local prefix. The remaining 64 bits are computed from the encapsulating header (e.g., 802.15.4 or IPv6 source address).

   b. If SAC=1, then SAM:

      - 00: 0 bits. The unspecified address (::).

      - 01: 64 bits. The address is derived using context information and the 64 bits carried in -line. Bits covered by context information are always used. Any IID bits not covered by context information are taken directly from the corresponding bits carried in-line.

- 10: 16 bits. The address is derived using context information and the 16 bits carried in -line. Bits covered by context information are always used. Any IID bits not covered by context information are taken directly from their corresponding bits in the 16-bit to IID mapping given by 0000:00ff:fe00:XXXX, where XXXX are the 16 bits carried in-line.

- 11: 0 bits. The address is fully elided and it is derived using context information and the encapsulating header (e.g., 802.15.4 or IPv6 source address). Bits covered by context information are always used. Any IID bits not covered by context information are computed from the encapsulating header.

The base format is two bytes (16 bits) long. If the CID (Context Identifier Extension) field has a value of 1, it means that an additional 8-bit Context Identifier Extension field immediately follows the Destination Address Mode (DAM) field. This would make the length be 24 bits or three bytes.

This additional octet identifies the pair of contexts to be used when the IPv6 source and/or destination address is compressed. The context identifier is 4 bits for each address, supporting up to 16 contexts. Context 0 is the default context. The two fields on the Context Identifier Extension are:

- SCI: Source Context Identifier. Identifies the prefix that is used when the IPv6 source address is statefully compressed.
- DCI: Destination Context Identifier. Identifies the prefix that is used when the IPv6 destination address is statefully compressed.

The Next Header field in the IPv6 header is treated in two different ways, depending on the values indicated in the NH (Next Header) field of the LOWPAN_IPHC encoding shown above.

If NH = 0, then this field is not compressed and all the 8 bits are carried in-line after the LOWPAN_IPHC.

If NH = 1, then the Next Header field is compressed and the next header is encoded using LOWPAN_NHC encoding. This results in the structure shown in the figure below.

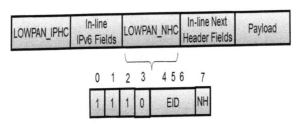

Figure 2.6. LoWPAN header

For IPv6 Extension headers the LOWPAN_NHC has the format shown in the figure, where:
• EID: IPv6 Extension Header ID:

  ○ 0: IPv6 Hop-by-Hop Options Header.

  ○ 1: IPv6 Routing Header.

  ○ 2: IPv6 Fragment Header.

  ○ 3: IPv6 Destination Options Header.

  ○ 4: IPv6 Mobility Header.

  ○ 5: Reserved.

  ○ 6: Reserved.

  ○ 7: IPv6 Header.

• NH: Next Header

  ○ 0: Full 8 bits for Next Header are carried in-line.

  ○ 1: Next Header field is elided and is encoded using LOWPAN_NHC. For the most part, the IPv6 Extension Header is carried unmodified in the bytes immediately following the LOWPAN_NHC octet.

NDP optimization

IEEE 802.15.4 and other similar link technologies have limited or no usage of multicast signalling due to energy conservation. In addition, the wireless network may not strictly follow the traditional concept of IP subnets and IP links. IPv6 Neighbor Discovery was not designed for non-transitive wireless links, since its reliance on the traditional IPv6 link concept and its heavy use of multicast make it inefficient and sometimes impractical in a low-power and lossy network.

For this reasons, some simple optimizations have been defined for IPv6 Neighbor Discovery, its addressing mechanisms and duplicate address detection for LoWPANs [RFC6775]:

1. Host-initiated interactions to allow for sleeping hosts.

2. Elimination of multicast-based address resolution for hosts.

3. A host address registration feature using a new option in unicast Neighbor Solicitation (NS) and Neighbor Advertisement (NA) messages.

4. A new Neighbor Discovery option to distribute 6LoWPAN header compression context to hosts.

5. Multi-hop distribution of prefix and 6LoWPAN header compression context.

6. Multi-hop Duplicate Address Detection (DAD), which uses two new ICMPv6 message types.

The two multi-hop items can be substituted by a routing protocol mechanism if that is desired. Three new ICMPv6 message options are defined:
1. The Address Registration Option (ARO).

2. The Authoritative Border Router Option (ABRO).

3. The 6LoWPAN Context Option (6CO)

Also two new ICMPv6 message types are defined:
1. The Duplicate Address Request (DAR).

2. The Duplicate Address Confirmation (DAC)

Chapter 4.
Embedded Devices

YOU HAVE AN idea for a *thing*, and you know that it has some sort of interactive or electronic side to it. What is the first step in turning that from a vision in your head into something in the real world?

You likely can try out a number of different parts of the behaviour in isolation, and that's a good starting point for your initial prototype. After you do some research on the Internet to find similar projects or look through the catalogues of component retailers, such as RS (www.rs-components.com/) or Farnell (www.farnell. com/), you'll have a list of possible components and modules which might let you achieve your goal.

The more you dabble in electronics and microcontrollers, the bigger your collection of spare parts and leftovers from previous projects will grow. When you sit down to try out some of your ideas, either you'll have rooted through your collection for parts which are close enough to those you identified in your research, or you'll have an assortment of freshly purchased components. Usually, it's a combination of the two.

That's the typical decider when first trying out an idea: you use what's easily to hand, partly because it's generally something you're familiar with already but also because it helps keep the costs down. Even if you know that the board you're using won't be the ideal fit for a final version, if it lets you try out some of the functionality more quickly or more cheaply, that can mean it's the right choice for now.

One of the main areas where something vastly overpowered, and in theory much more expensive, can be the right choice for prototyping is using a mobile phone, laptop, or desktop computer to develop the initial software. If you already have a phone or computer which you can use, using it for your prototype isn't actually any more expensive.

However, if you haven't been playing around with electronics already and don't have a collection of development boards gathering dust in your desk drawer, how do you choose which one to buy? In this chapter, we explain some of the differences and features of a number of popular options. Over time the list will change, but you should still be able to work out how the same criteria we discussed in the preceding chapter apply to whichever boards you are considering.

This chapter starts with a look at electronics because whatever platform you end up choosing, the rest of the circuitry that you will build to connect it to will be pretty much the same. Then we choose four different platforms that you could use as a basis for your Internet of Things prototype. They aren't the only options, but they cover the breadth of options available. By the end of the chapter you will have a good feel for the trade-offs between the different options and enough knowledge of the example boards to make a choice on which to explore further.

ELECTRONICS

Before we get stuck into the ins and outs of microcontroller and embedded computer boards, let's address some of the electronics components that you might want to connect to them.

Don't worry if you're scared of things such as having to learn soldering. You are unlikely to need it for your initial experiments. Most of the prototyping can be done on what are called *solderless breadboards*. They enable you to build components together into a circuit with just a push-fit connection, which also means you can experiment with different options quickly and easily.

When it comes to thinking about the electronics, it's useful to split them into two main categories:

Sensors: Sensors are the ways of getting information *into* your device, finding out things about your surroundings.

Actuators: Actuators are the *outputs* for the device—the motors, lights, and so on, which let your device do something to the outside world.

Within both categories, the electronic components can talk to the computer in a number of ways.

The simplest is through digital I/O, which has only two states: a button can either be pressed or not; or an LED can be on or off. These states are usually connected via general-purpose input/output (GPIO) pins and map a digital 0 in the processor to 0 volts in the circuit and the digital 1 to a set voltage, usually the voltage that the processor is using to run (commonly 5V or 3.3V).

If you want a more nuanced connection than just on/off, you need an analogue signal. For example, if you wire up a potentiometer to let you read in the position of a rotary knob, you will get a varying voltage, depending on the knob's location. Similarly, if you want to run a motor at a speed other than off or full-speed, you need to feed it with a voltage somewhere between 0V and its maximum rating.

Because computers are purely digital devices, you need a way to translate between the analogue voltages in the real world and the digital of the computer.

An analogue-to-digital converter (ADC) lets you measure varying voltages. Microcontrollers often have a number of these converters built in. They will convert the voltage level between 0V and a predefined maximum (often the same 5V or 3.3V the processor is running at, but sometimes a fixed value such as 1V) into a number, depending on the accuracy of the ADC. The Arduino has 10-bit ADCs, which by default measure voltages between 0 and 5V. A voltage of 0 will give a reading of 0; a voltage of 5V would read 1023 (the maximum value that can be stored in a 10-bits); and voltages in between result in readings relative to the voltage. 1V would map to 205; a reading of 512 would mean the voltage was 2.5V; and so on.

The flipside of an ADC is a DAC, or digital-to-analogue converter. DACs let you generate varying voltages from a digital value but are less common as a standard feature of microcontrollers. This is due to a technique called *pulse-width modulation* (PWM), which gives an approximation to a DAC by rapidly turning a digital signal on and off so that the average value is the level you desire. PWM requires simpler circuitry, and for certain applica-tions, such as fading an LED, it is actually the preferred option.

For more complicated sensors and modules, there are interfaces such as Serial Peripheral

Interface (SPI) bus and Inter-Integrated Circuit (I2C). These standardized mechanisms allow modules to communicate, so sensors or things such as Ethernet modules or SD cards can interface to the microcontroller.

Naturally, we can't cover all the possible sensors and actuators available, but we list some of the more common ones here to give a flavour of what is possible.

SENSORS

Pushbuttons and switches, which are probably the simplest sensors, allow some user input. Potentiometers (both rotary and linear) and rotary encoders enable you to measure movement.

Sensing the environment is another easy option. Light-dependent resistors (LDRs) allow measurement of ambient light levels, thermistors and other temperature sensors allow you to know how warm it is, and sensors to measure humidity or moisture levels are easy to build.

Microphones obviously let you monitor sounds and audio, but piezo elements (used in certain types of microphones) can also be used to respond to vibration.

Distance-sensing modules, which work by bouncing either an infrared or ultrasonic signal off objects, are readily available and as easy to interface to as a potentiometer.

ACTUATORS

One of the simplest and yet most useful actuators is light, because it is easy to create electronically and gives an obvious output. Light-emitting diodes (LEDs) typically come in red and green but also white and other colours.
RGB LEDs have a more complicated setup but allow you to mix the levels of red, green, and blue to make whatever colour of light you want. More complicated visual outputs also are available, such as LCD screens to display text or even simple graphics.
Piezo elements, as well as *responding* to vibration, can be used to *create* it, so you can use a piezo buzzer to create simple sounds and music. Alternatively, you can wire up outputs to speakers to create more complicated synthesized sounds.
Of course, for many tasks, you might also want to use components that *move* things in the real world. Solenoids can by used to create a single, sharp pushing motion, which could be useful for pushing a ball off a ledge or tapping a surface to make a musical sound.
More complicated again are motors. Stepper motors can be moved in *steps*, as the name implies. Usually, a fixed number of steps perform a full rotation. DC motors simply move at a given speed when told to. Both types of motor can be one-directional or move in both directions. Alternatively, if you want a motor that will turn to a given angle, you would need a servo. Although a servo is more controllable, it tends to have a shorter range of motion, often 180 or fewer degrees (whereas steppers and DC motors turn indefinitely). For all the kinds of motors that we've mentioned, you typically want to connect the motors to gears to alter the range of motion or convert circular movement to linear, and so on.

If you want to dig further into the ways of interfacing your computer or microcontroller with the real world, the "Interfacing with Hardware" page on the Arduino Playground website (http://playground.arduino.cc//Main/ InterfacingWithHardware) is a good place to start. Although Arduino-focused, most of the suggestions will translate to other platforms with minimal changes. For a more in-depth introduction to electronics, we recommend

Electronics For Dummies *(Wiley, 2009).*

SCALING UP THE ELECTRONICS

From the perspective of the electronics, the starting point for prototyping is usually a "breadboard". This lets you push-fit components and wires to make up circuits without requiring any soldering and therefore makes experimen-tation easy. When you're happy with how things are wired up, it's common to solder the components onto some protoboard, which may be sufficient to make the circuit more permanent and prevent wires from going astray.

Moving beyond the protoboard option tends to involve learning how to lay out a PCB. This task isn't as difficult as it sounds, for simple circuits at least, and mainly involves learning how to use a new piece of software and understanding some new terminology.

For small production runs, you'll likely use through-hole components, so called because the legs of the component go through holes in the PCB and tend to be soldered by hand. You will often create your designs as companion boards to an existing microcontroller platform—generally called *shields* in the Arduino community. This approach lets you bootstrap production without worrying about designing the entire system from scratch.

Journey to a Circuit Board

Let's look at the evolution of part of the Bubblino circuitry, from initial testing, through prototype, to finished PCB:

The first step in creating your circuit is generally to build it up on a breadboard. This way, you can easily reconfigure things as you decide exactly how it should be laid out.

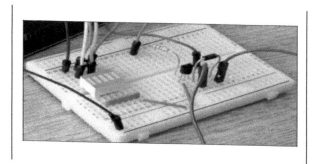

The breadboard.

When you are happy with how the circuit works, soldering it onto a stripboard will make the layout permanent. This means you can stop worrying about one of the wires coming loose, and if you're going to make only one copy of the circuit, that might be as far as you need take things.

When you want to scale things even further, moving to a combined board allows you to remove any unnecessary components from the microcon-troller board, and switching to surface mount components—where the legs of the chips are soldered onto the same surface as the chip—eases the board's assembly with automated manufacturing lines.

PCB design and the options for manufacturing are covered in much greater detail in Chapter 10, "Moving to Manufacture".

EMBEDDED COMPUTING BASICS

The rest of this chapter examines a number of different embedded comput-ing platforms, so it makes sense to first cover some of the concepts and terms that you will encounter along the way.

Providing background is especially important because many of you may have little or no idea about what a microcontroller is. Although we've been talking about computing power getting cheaper and more powerful, you cannot just throw a bunch of PC components into something and call it an Internet of Things product. If you've ever opened up a desktop PC, you've seen that it's a collection of discrete modules to provide different aspects of functionality. It has a main motherboard with its processor, one or two smaller circuit boards providing the RAM, and a hard disk to provide the long-term storage. So, it has a lot of components, which provide a variety of general-purpose functionality and which all take up a corresponding chunk of physical space.

MICROCONTROLLERS

Internet of Things devices take advantage of more tightly integrated and miniaturized solutions—from the most basic level of microcontrollers to more powerful system-on-chip (SoC) modules. These systems combine the processor, RAM, and storage onto a single chip, which means they are much more specialized, smaller than their PC equivalents, and also easier to build into a custom design.

These microcontrollers are the engines of countless sensors and automated factory machinery. They are the last bastions of 8-bit computing in a world that's long since moved to 32-bit and beyond. Microcontrollers are very limited in their capabilities—which is why 8-bit microcontrollers are still in use, although the price of 32-bit microcontrollers is now dropping to the level where they're starting to be edged out. Usually, they offer RAM capabilities measured in kilobytes and storage in the tens of kilobytes. However, they can still achieve a lot despite their limitations.

You'd be forgiven if the mention of 8-bit computing and RAM measured in kilobytes gives you flashbacks to the early home computers of the 1980s such as the Commodore 64 or the Sinclair ZX Spectrum. The 8-bit microcon-trollers have the same sort of internal workings and similar levels of memory to work with. There have been some improvements in the intervening years, though—the modern chips are much smaller, require less power, and run about five times faster than their 1980s counterparts.

Unlike the market for desktop computer processors, which is dominated by two manufacturers

(Intel and AMD), the microcontroller market consists of many manufacturers. A better comparison is with the automotive market. In the same way that there are many different car manufacturers, each with a range of models for different uses, so there are lots of microcontroller manufacturers (Atmel, Microchip, NXP, Texas Instruments, to name a few), each with a range of chips for different applications.

The ubiquitous Arduino platform is based around Atmel's AVR ATmega family of microcontroller chips. The on-board inclusion of an assortment of GPIO pins and ADC circuitry means that microcontrollers are easy to wire up to all manner of sensors, lights, and motors. Because the devices using them are focused on performing one task, they can dispense with most of what we would term an operating system, resulting in a simpler and much slimmer code footprint than that of a SoC or PC solution.

In these systems, functions which require greater resource levels are usually provided by additional single-purpose chips which at times are more powerful than their controlling microcontroller. For example, the WizNet Ethernet chip used by the Arduino Ethernet has eight times more RAM than the Arduino itself.

SYSTEM-ON-CHIPS

In between the low-end microcontroller and a full-blown PC sits the SoC (for example, the BeagleBone or the Raspberry Pi). Like the microcontroller, these SoCs combine a processor and a number of peripherals onto a single chip but usually have more capabilities. The processors usually range from a few hundred megahertz, nudging into the gigahertz for top-end solutions, and include RAM measured in megabytes rather than kilobytes. Storage for SoC modules tends not to be included on the chip, with SD cards being a popular solution.

The greater capabilities of SoC mean that they need some sort of operating system to marshal their resources. A wide selection of embedded operating systems, both closed and open source, is available and from both specialised embedded providers and the big OS players, such as Microsoft and Linux.

Again, as the price falls for increased computing power, the popularity and familiarity of options such as Linux are driving its wider adoption.

CHOOSING YOUR PLATFORM

How to choose the *right* platform for your Internet of Things device is as easy a question to answer as working out the meaning of life. This isn't to say that it's an impossible question—more that there are almost as many answers as there are possible devices. The platform you choose depends on the particular blend of price, performance, and capabilities that suit what you're trying to achieve. And just because you settle on one solution, that doesn't mean somebody else wouldn't have chosen a completely different set of options to solve the same problem.

Start by choosing a platform to prototype in. The following sections discuss some of the factors that you need to weigh—and possibly play off against each other—when deciding how to build your device.

Processor Speed

The processor speed, or clock speed, of your processor tells you how fast it can process the individual instructions in the machine code for the program it's running. Naturally, a faster processor speed means that it can execute instructions more quickly.

The clock speed is still the simplest proxy for raw computing power, but it isn't the only one. You might also make a comparison based on millions of instructions per second (MIPS), depending on what numbers are being reported in the datasheet or specification for the platforms you are comparing.

Some processors may lack hardware support for floating-point calculations, so if the code involves a lot of complicated mathematics, a by-the-numbers slower processor with hardware floating-point support could be faster than a slightly higher performance processor without it.

Generally, you will use the processor speed as one of a number of factors when weighing up similar systems. Microcontrollers tend to be clocked at speeds in the tens of MHz, whereas SoCs run at hundreds of MHz or possibly low GHz.

If your project doesn't require heavyweight processing—for example, if it needs only networking and fairly basic sensing—then some sort of micro-controller will be fast enough. If your device will be crunching lots of data—for example, processing video in real time—then you'll be looking at a SoC platform.

RAM

RAM provides the working memory for the system. If you have more RAM, you may be able to do more things or have more flexibility over your choice of coding algorithm. If you're handling large datasets on the device, that could govern how much space you need. You can often find ways to work around memory limitations, either in code (see Chapter 8, "Techniques for Writing Embedded Code") or by handing off processing to an online service (see Chapter 7, "Prototyping Online Components").

It is difficult to give exact guidelines to the amount of RAM you will need, as it will vary from project to project. However, microcontrollers with less than 1KB of RAM are unlikely to be of interest, and if you want to run standard encryption protocols, you will need at least 4KB, and preferably more.

For SoC boards, particularly if you plan to run Linux as the operating system, we recommend at least 256MB.

Networking

How your device connects to the rest of the world is a key consideration for Internet of Things products. Wired Ethernet is often the simplest for the user—generally plug and play—and cheapest, but it requires a physical cable. Wireless solutions obviously avoid that requirement but introduce a more complicated configuration.

WiFi is the most widely deployed to provide an existing infrastructure for connections, but it can be more expensive and less optimized for power consumption than some of its competitors.

Other short-range wireless can offer better power-consumption profiles or costs than WiFi but usually with the trade-off of lower bandwidth. ZigBee is one such technology, aimed particularly at sensor networks and scenarios such as home automation. The recent Bluetooth LE protocol (also known as Bluetooth 4.0) has a very low power-consumption profile similar to ZigBee's and could see more rapid adoption due to its inclusion into standard Bluetooth chips included in phones and laptops. There is, of course, the existing Bluetooth standard as another possible choice. And at the boring-but-very-cheap end of the market sit long-established options such as RFM12B which operate in the 434 MHz radio spectrum, rather than the 2.4 GHz range of the other options we've discussed.

For remote or outdoor deployment, little beats simply using the mobile phone networks. For low-bandwidth, higher-latency communication, you could use something as basic as SMS; for higher data rates, you will use the same data connections, like 3G, as a smartphone.

USB

If your device can rely on a more powerful computer being nearby, tethering to it via USB can be an easy way to provide both power and networking. You can buy some of the microcontrollers in versions which include support for USB, so choosing one of them reduces the need for an extra chip in your circuit.

Instead of the microcontroller presenting itself as a device, some can also act as the USB "host". This configuration lets you connect items that would normally expect to be connected to a computer—devices such as phones, for example, using the Android ADK, additional storage capacity, or WiFi dongles.

Devices such as WiFi dongles often depend on additional software on the host system, such as networking stacks, and so are better suited to the more computer-like option of SoC.

Power Consumption

Faster processors are often more power hungry than slower ones. For devices which might be portable or rely on an unconventional power supply (batteries, solar power) depending on where they are installed, power consumption may be an issue. Even with access to mains electricity, the power consumption may be something to consider because lower consumption may be a desirable feature.

However, processors may have a minimal power-consumption sleep mode. This mode may allow you to use a faster processor to quickly perform operations and then return to low-power sleep. Therefore, a more powerful processor may *not* be a disadvantage even in a low-power embedded device.

Interfacing with Sensors and Other Circuitry

In addition to talking to the Internet, your device needs to interact with something else—either sensors to gather data about its environment; or motors, LEDs, screens, and so on, to provide output. You could connect to the circuitry through some sort of peripheral bus—SPI and I2C being common ones—or through ADC or DAC modules to read or write varying voltages; or through generic GPIO pins, which provide digital on/off inputs or outputs. Different microcontrollers or SoC solutions offer different mixtures of these interfaces in differing numbers.

Physical Size and Form Factor

The continual improvement in manufacturing techniques for silicon chips means that we've long passed the point where the limiting factor in the size of a chip is the amount of space required for all the transistors and other components that make up the circuitry on the silicon. Nowadays, the size is governed by the number of connections it needs to make to the surrounding components on the PCB. With the traditional through-hole design, most commonly used for home-made circuits, the legs of the chip are usually spaced at 0.1" intervals. Even if your chip has relatively few connections to the surrounding circuit—16 pins is nothing for such a chip—you will end up with over 1.5" (~4cm) for the perimeter of your chip. More complex chips can easily run to over a hundred connections; finding room for a chip with a 10" (25cm) perimeter might be a bit tricky!

You can pack the legs closer together with surface-mount technology because it doesn't require holes to be drilled in the board for connections. Combining that with the trick of hiding some of the connections on the underside of the chip means that it is possible to use the complex designs without resorting to PCBs the size of a table.

The limit to the size that each connection can be reduced to is then governed by the capabilities and tolerances of your manufacturing process. Some surface-mount designs are big enough for home-etched PCBs and can be hand-soldered. Others require professionally produced PCBs and accurate pick-and-place machines to locate them correctly.

Due to these trade-offs in size versus manufacturing complexity, many chip designs are available in a number of different form factors, known as *packages*. This lets the circuit designer choose the form that best suits his particular application.

All three chips pictured in the following figure provide identical functional-ity because they are all AVR ATmega328 microcontrollers. The one on the left is the through-hole package, mounted here in a socket so that it can be swapped out without soldering. The two others are surface mount, in two different packages, showing the reduction in size but at the expense of ease of soldering.

Through-hole versus surface-mount ATmega328 chips.

Looking at the ATmega328 leads us nicely into comparing some specific embedded computing platforms. We can start with a look at one which so popularized the ATmega328 that a couple of years ago it led to a worldwide shortage of the chip in the through-hole package, as for a short period demand outstripped supply.

Chapter 5.
Cloud Databases for the Internet of Things

The phrase "Internet of Things" started life in 1999 by Kevin Ashton, co-founder and executive director of Auto-ID Center [55].

"Internet of Things (IoT) is an integrated part of Future Internet and could be de ned as a dynamic global network infrastructure with self con guring capabilities based on standard and interoperable communication protocols where physical and virtual things have identities, physical attributes, and virtual personalities and use intelligent interfaces, and are seamlessly integrated into the information network."

To make it simpler, IoT refers to a world of physical and virtual objects (things) which are uniquely identi ed and capable of interacting with each other, with people, and with the environment. It allows people and things to be connected at anytime and anyplace, with anything and anyone. Communication among the things is achieved by exchanging the data and information sensed and generated during their interactions.

Internet of Things vision

The broad future vision of IoT is to make the things able to react to physical events with suitable behavior, to understand and adapt to their environment, to learn from, collaborate with and manage other things, and all these are autonomous with or without direct human intervention. To achieve such a goal, numerous researches have been carried out, which emphasize on di erent aspects of the IoT. The followings are the three main concrete visions of the IoT that most of the researches are focusing on [8] [12]:

Things-oriented Vision

Originally, the IoT started with the development of RFID (Radio Frequency Identi cation) tagged objects that communicate over the Internet. RFID along with the Electronic Product Code (EPC) global framework [57] is one of the key components of the IoT architecture. The technology targets a global EPC system of RFID tags that provide object identi cation and traceability.

However, the vision is not limited to RFID. Many other technologies are involved in the things-vision of IoT, including Universally Unique IDenti er (UUID) [38], Near Field Communications (NFC) [65], and Wireless Sensor and Actuator Networks [61]. Those in conjunction with RFID are to be the core components that make up the Internet of Things. Applying these technologies, the concept of things has been expanded to be of any kind: from human to electronic devices such as computers, sensors, actuators, phones. In fact, any everyday object might be made smart and become a thing in the network. For example, TVs, vehicles, books, clothes, medicines, or food can be equipped with embedded sensor devices that make them uniquely addressable, be able to collect information, connect to the Internet, and build a network of networks of IoT objects.

Internet-oriented Vision

A focus of the Internet-oriented vision is on the IP for Smart Objects (IPSO)

which proposes to use the Internet Protocol to support smart objects connection around the world. As a result, this vision poses the challenge of developing the Internet infrastructure with an IP address space that can accommodate the huge number of connecting things. The development of IPv6 has been recognized as a direction to deal with the issue.

Another focus of this vision is the development of the Web of Things [30], in which the Web standards and protocols are used to connect embedded devices installed on everyday objects. That is to make use of the current popular standards such as URI, HTTP, RESTful API to access physical devices, and integrate those objects into the Web.

Semantic-oriented Vision

The heterogeneity of IoT things along with the huge number of objects involved impose a signi cant challenge for the interoperability among them. Semantic technologies [13] have shown potential for a solution to represent, exchange, integrate, and manage information in a way that conforms with the global nature of the Internet of Things. The idea is to create a standardized description for heterogeneous resources, develop comprehensive shared information models, provide semantic mediators and execution environments [52], thus accommodating semantic interoperability and integration for data coming from various sources.

Internet of Things data

With its powerful ability, the scope of the Internet of Things is wide. It can provide applicability and pro ts for users and organizations in a variety of elds, including environmental monitoring, inventory and product management, customer pro ling, market research, health care, smart homes, or security and surveillance [42]. For instance, digital billboards use face recognition to analyze passing shoppers, identify their gender and age range, and change the advertisement content accordingly. A smart refrigerator keeps track of food items' availability and expiry date, then autonomously orders new ones if needed. A sensor network used to monitor crop conditions can control farming equipments to spray fertilizer on areas that are lack of nutrients. Examples for such applications of IoT are countless. Therefore, the types of data transmitted in the Internet of Things are also unlimited. It could be either discrete or continuous, input by humans or auto-generated. Generally, IoT data include, but not limited to, the following categories [20][17].

RFID Data. Radio Frequency Identi cation [64] systems are said to be a main component of the IoT [12]. The technique uses radio wave for identi cation and tracking purposes. An RFID tagging system includes several RFID tags that are uniquely identi ed and can be attached to everyday objects. The tag can store information internally and transmit data as radio waves to an RFID reader through an antenna. Hence, the technology can be used to monitor objects in real time. For example, it can replace bar codes in supply chain management, stock control, or used to track livestock and wildlife. In healthcare, VeriChip [29] is an RFID tag that can be injected under human's skin. It is used to biometrically identify patients and provide critical information about their medical records.

Sensor Data. Sensor networks [9] have been widely spread nowadays from small to large scale. They are also a key component in the Internet of Things. Their usage varies from recording and monitoring environment parameters or patient conditions in real time to tracking customer behavior and other applications. Several common parameters are temperature, power, humidity, electricity, sound, blood pressure, and heart rate. Data format can also be di erent, from numeric or text based to multimedia data. For this data type, the normal question is how often the data is to be captured, whether continuously, periodically, or when queried. In any case, the result could be an enormous volume of data, which in turn raises a challenge of storage as well as how to do querying, data mining, and data analysis on such an amount of data with a real-time demand. Additionally, sensor data generation tends to be continuous. As time goes by, some data become old and less valuable. Hence, the system is responsible to decide which data to keep, when to remove or archive old data, or how to distribute new data to active data warehouses used for frequent querying.

In the book, one of our focus is on sensor scalar data. The context for the sensor data benchmark is based on the Home Energy Management System (HEMS) developed by There corporation [2]. The system uses smart metering sensors to monitor the electric energy consumption of households. The energy is periodically measured and recorded data are sent to a central database. Customers can then get the real time report about the energy usage in their house via a provided web service.

Multimedia Data. The term refers to the convergence of text, picture, audio, and video into a single form. Multimedia data nds its application in numerous areas including surveillance, entertainment, journalism, advertisement, education and more. As a result, it can easily contribute a large source of data to the Internet of Things.

Positional Data. This data represents the location of an object within a positioning system, for example a global positioning system (GPS). Positional data is highly relevant in the work of mobile computing where objects are either static or mobile, or geographical information system.

Descriptive Data and Metadata about Objects (or Processes and Systems). This kind of data describes the attributes of a certain object, to help identify the object type, to address the object, and to di erentiate it with other objects. For example, an IoT object might have data \TV", \Samsung", \40 inches" and the corresponding metadata for it are \Type", \Brand", \Size".

Command Data. Some of the data coming into the network will be command data which are used to control devices such as actuators. The interfaces of each system are di erent, and so the format of command data will be di erent as well.

Cloud Databases

By its name, a cloud database is a database that runs on a cloud computing platform, such as Amazon Web Services, Rackspace and Microsoft Azure. The cloud platform can provide databases as a specialized service, or provide virtual machines to deploy any databases on. Cloud databases could be either relational or non-relational databases. Compared to local databases, cloud databases are guaranteed higher scalability as well as availability and stability. Thanks to the elasticity of cloud computing, hardware and software resources can be added to and removed from the cloud without much e ort. Users only need to pay for the consumed resource while the expenses for physical servers, networking equipments, infrastructure maintenance and administration are shared among clients, thus reducing the overall cost. Additionally, database service is normally provided along with automated features such as backup and recovery, failover, on-the-go scaling, and load balancing.

Amazon Web Services

The most prominent cloud computing provider these days is Amazon with its Amazon Web Services (AWS) [4]. Clients can purchase a database service from a set of choices:

Amazon RDS. Amazon Relational Database Service is used to build up a relational database system in the cloud with high scalability and little administration e ort. The service comes with a choice of the three popular SQL databases including MySQL, Oracle, and Microsoft SQL Server.

Amazon DynamoDB, Amazon SimpleDB. These are the key-value NoSQL databases provided by Amazon. The administrative work here is also minimal. DynamoDB o ers very high performance and scalability but simple query capability. Meanwhile, SimpleDB is suitable for a smaller data set that requires query exibility, but with a limitation on storage (10GB) and request capacity (normally 25 writes/second).

Amazon S3. The Simple Storage Service provides a simple web service interface (REST or SOAP) to store and retrieve unstructured blobs of data, each is up to 5 TB size and has a unique key. Therefore, it is suitable for storing large objects or data that is not accessed frequently.

Amazon EC2 (Amazon Elastic Compute Cloud). When clients require a particular database or full administrative control over their databases, the database can be deployed on an Amazon EC2 instance, and their data can be stored temporarily on an Amazon EC2 Instance Store or persistently on an Amazon Elastic Block Store (Amazon EBS) volume.

Scalability

Scalability is one key point of cloud databases that make them more advantageous and suitable for large systems than local databases. Scalability is the ability of a system to expand to handle load increases. The dramatic growth in data volumes and the demand to process more data in a shorter time are putting a pressure on current database systems. The question is to nd a cost-e ective solution for scalability, which is essential for cloud computing and large-scale Web sites such as Facebook, Amazon, or eBay. Scalability can be achieved by either scaling vertically or horizontally [48]. Vertical scaling (scale up) means to use a more powerful machine by adding processors and storage. This way of scaling can only go to a certain extent. To get beyond that extent, horizontal scaling (scale out) should be used. That is to use a cluster of multiple independent servers to increase processing power.

Currently, there are two methods that can be used to achieve horizontal scalability, that is, replication and sharding.

Replication

Replication is the process of copying data to more than one server. It increases the robustness of the system by reducing the risk of data loss and one single point of failure. The nodes can be distributed closer to clients, thus reducing latency in some cases, but also making the nodes far away from each other lengthens the data propagating process. Besides, replication can e ectively improve read performance as read queries are spread across multiple nodes. However, write performance normally decreases as data have to be written to multiple nodes. Depending on the database system, replication can be synchronous, asynchronous, or semi-synchronous [53]. A database using synchronous replication only returns a write call when it has nished on the slaves (usually a majority of the slaves) and received their acknowledgements. On the other hand, in asynchronous replication, a write is considered complete as soon as the data is written on the master while there might be a lag in updating the slaves.

Replication can either be master-slave or multi-master. In case of master-slave replication, one single node of the cluster is designated as the master, and data modi cations can only be performed on that node. Allowing one single master makes it easier to ensure system consistency, and that node can be dedicated to write operations while the others (slaves) take care of read operations. Meanwhile, multi-master replication is more exible as all nodes can receive write calls from clients, but they are responsible for resolving con icts during data synchronization.

Replication can provide some useful features, such as automatic load balancing (for example on a round-robin basis), or failover which is the ability to automatically switch from a primary system component (for example server, database) to a secondary one in case of a sudden failure.

Sharding

In short, sharding [19] means horizontal partitioning data across a number of servers. One database (or one table) is divided into smaller ones, all have the same or similar structure. Each partition is called a shard. Partitioning is done with a \shared-nothing" approach, that is, the servers are CPU, memory and disk independent. Hence, sharding solution is needed for systems that have data sets larger than the storage capacity of a single node, or systems that are write-intensive in which one node cannot write data fast enough to meet the demand.

Scalability by sharding is achieved through the distribution of processing (both reads and writes) across multiple shards. A smaller data set can also outperform a large one. Moreover, sharding is cost-e ective as it is possible to use commodity hardware rather than an expensive high-end multi-core server.

On the other hand, sharding also poses several challenges. First and foremost is choosing an e ective sharding strategy, since using a wrong one can actually inhibit performance. A database table can be divided in many ways, for example, based on the value ranges of one or several elds that appear in all data items, or using a hash function to perform on an item eld. The ideal option is the one that can distribute data and load evenly, take advantage of distributed processing while avoiding cross-shard joins. However, which solution to choose highly depends on the query orientation, data structures, and key distribution nature of the system. At the same time, sharding increases the complexity of a system. The system highly relies on its coordinating and rebalancing functionalities. Scattered data complicates the process of management, backup, and maintaining data integrity, especially when there is a change in the data schema. Besides, partitioning data causes single points of failure. Corruption of one shard due to network or hardware can lead to a failure of the entire table. To avoid this, large systems usually apply a combination of sharding and replication, where each shard is a replicated set of nodes.

Database characteristics of SQL vs. NoSQL

A lot of work has been done on studying the characteristics and features of di erent kinds of databases. Many reviews and surveys comparing SQL versus NoSQL as well as comparing multiple NoSQL databases are available [58][31][45].

Padhy et al. [46] characterized the three main types of NoSQL databases, that is, key-value, column-oriented, and document stores. Simultaneously, the authors gave detailed description about the data model and architecture of several popular databases, namely, Amazon SimpleDB,

CouchDB, Google Big Table, Cassandra, MongoDB, and HBase.

Hecht et al. [32] evaluated the four NoSQL database classes, the three above along with the graph databases. The underlying technologies were compared from di erent aspects, from data models, queries, concurrency controls, to scalability, but all were evaluated under the consideration of the database applicability for systems of di erent requirements.

Meanwhile, Jatana1 et al. [34] studied the two broad categories of databases: relational and non -relational. The authors gave an overview of each database class, along with their advantages and disadvantages. Several widely used databases were also brie y introduced. Finally, the paper highlighted the key di erences between the two classes of database.

Database performance of SQL vs. NoSQL

As regards database performance measurement, countless tests have been run, but most are individual, small scale, or case speci c local tests. Nevertheless, there are several researches have been carried out in an attempt to demonstrate the performance with real world loads.

Datastax Corporation examined three NoSQL databases: key-value store Apache Cassandra, column oriented Apache HBase, and document store MongoDB on Amazon EC2 m1 extra large instances [22]. The results showed that Cassandra outperformed the other two by a large margin, while MongoDB was the worst. However, there was no SQL database involved, nor was there any other document database to compare with MongoDB, which is one of the main focus in our tests.

In the paper written by Konstantinou et al. [35], the elasticity of NoSQL databases, including HBase, Cassandra, and Riak, was veri ed and compared as the authors examined the changes in query throughput when the server cluster size changed. The results showed HBase as the fastest and most scalable when the system was read intensive, whereas Cassandra performed and scaled well in a write intensive environment, and nodes could be added without a transitional delay time. Apart from that, the authors proposed a prototype for an automatic cluster resize module that could t the system requirements.

Meanwhile, Rabl et al. [49] addressed the challenge of storing application performance management data, and analyzed the scalability and performance of six databases including MySQL and ve NoSQL databases. The benchmark showed the latency and throughput of those databases under di erent workload test cases. Again Cassandra was the clear winner throughout the experiments, while HBase got the lowest throughput. When it comes to sharding, MySQL achieved nearly as high throughput as Cassandra. Although a standalone Redis outperformed the others when the system was read intensive, its performance of a sharded implementation dropped with an increasing number of nodes. The same case applied for VoltDB in a sharded system, thus Redis and VoltDB did not scale very well.

Tudorica et al. [58] compared MySQL, Cassandra, HBase, and Sherpa. The experiments concluded that the SQL database was not as e cient as the NoSQL ones when it comes to data of massive volume, especially on write intensive systems. However, MySQL could have relatively high performance on read intensive systems.

Internet of Things storage

As regards more types of data related to IoT data, van der Veen et al. [59] compared PostgreSQL, Cassandra, and MongoDB as a storage for sensor data. The tests were not run in a cloud environment, but run in a comparison between a physical server and a virtual machine. The paper is closely related to our work for the similar used data structure. The result did not show a solely winner as MongoDB won at single writes and PostgreSQL at multiple reads. The impact of virtualization was unclear for it was different in each case.

Other solutions for storing IoT data have also been proposed. One is a storage management solution called IOTMDB which is based on NoSQL [40]. The system came with the strategies for a common IoT data expression in the form of key-value, as well as a data preprocessing and sharing mechanism.

Pintus et al. [47] introduced a system called Paraimpu, a social scalable Web-based platform that allowed clients to connect, use, share data and functionalities, and build a Web of Things connecting HTTP-enabled smart devices like sensors, actuators with virtual things such as services, social networks, and APIs. The platform uses MongoDB as the database server, provides models and interfaces that help to abstract and adopt different kinds of things and data.

Another solution is the SeaCloudDM [25], a cloud data management framework for sensor data. The solution addressed the challenges that the data are dynamic, various, massive, and spatial-temporal (i.e., each data sampling corresponds to a specific time and location). To provide a uniformed storage mechanism for heterogeneous sensor sampling data, the system combined the use of the relational model and the key-value model, and was implemented with PostgreSQL database. Its multi-layer architecture was claimed to reduce the amount of data to be processed at the cloud management layer. Besides, the paper also came with several experiments that showed a promising result for the performance of the system when storing and querying a huge volume of data.

Meanwhile, Di Francesco et al.[24] proposed a document-oriented data model and storage infrastructure for IoT heterogeneous as well as multimedia data. The system used CouchDB as the database server, taking advantage of its RESTful API, and supporting other features such as replication, batch processing, and change notifications. The authors also provided an optimized document uploading scheme for multimedia data that showed a clear enhancement in performance.

Chapter 6.
IoT & Cities

Cities around the world have been the locus of innovation in the use of the Internet of Things. Through "smart city" initiatives and entrepreneurship, cities are experimenting with IoT applications to improve services, relieve traffic congestion, conserve water and energy, and improve quality of life. Large, concentrated populations and complex infrastructure make cities a target-rich environment for IoT applications. And cities have the most to gain: they are engines of global economic growth, and with urbanization in developing economies, 60 percent of the world's population—about 4.7 billion people—will live in cities in 2025.

We found the highest economic impact of IoT applications we analyzed to be concentrated in a few use areas, such as public health and safety, transportation, and resource management. IoT applications in public safety and health include air and water quality monitoring. Transportation applications range from traffic-control systems to smart parking meters to autonomous vehicles. Resource and infrastructure management uses include sensors and smart meters to better manage water and electric infrastructure.

Based on current adoption and likely growth rates, we estimate that the economic impact of the Internet of Things in cities (for the applications we size) could be $930 billion to $1.7 trillion globally in 2025. Our estimates of potential impact are based on the value of improved health and safety (automobile deaths avoided and reduction in pollution-related illnesses, for example), the value of time saved through IoT applications, and more efficient use of resources. We estimate the economic impact of illness and deaths avoided by IoT applications using quality-adjusted life years, a measure of the economic value of a year of perfect health in a particular economy. There are many additional social and environmental benefits, such as tracking lost children and higher social engagement, which we do not attempt to size.

Reaching this level of impact depends on addressing the technical and regulatory issues common to other settings—the need for lower-cost hardware and building protections for privacy and security. In cities, there also will need to be political consensus to support IoT applications, which in many cases will require investment of public funds. For example, the decision to install adaptive traffic-control systems that adjust traffic lights using sensor data will require motivated city authorities and politicians.

Definition

We define the city setting to include all urban settlements, consistent with the definition used by the United Nations in its World Urbanization Prospects report. In our estimates of IoT impact in cities, we do not include applications used in homes or the use of IoT devices for health and fitness, which are counted in the home and human settings. Exhibit 24 illustrates some of the emerging applications of IoT technology that can improve the performance of resource management, transportation, and public safety and health.

Potential for economic impact

For the applications we sized, we estimate that the potential economic impact of the Internet of Things in the city setting could exceed $1.7 trillion per year in 2025. The single most economically important application could be public health, where we estimate that nearly $700 billion a year of value might be captured in 2025 from the improved health outcomes that would result from water and air monitoring. Taken together, however, transportation applications could have even larger economic impact—more than $800 billion per year in 2025. Traffic applications, including real-time traffic flow management, smart meters and more efficient use of public transportation (reducing wait times by using real-time bus and train information), could be worth more than $570 billion a year globally. Autonomous vehicles could contribute as much as $235 billion by reducing traffic accidents, fuel consumption, and carbon emissions.

Public health and safety

IoT technology has several applications in urban public health and safety, which could have an economic impact of about $700 billion per year in 2015. These applications include using video cameras for crime monitoring, improving emergency services and disaster response, and monitoring street lights and the structural health of buildings. The biggest impact, however, would come from the application of IoT technology in air and water quality monitoring.

Air and water quality monitoring

The World Health Organization estimates that 3.7 million deaths were linked to outdoor air pollution in 2012, with the bulk of deaths occurring in low- and middle-income countries. IoT technology provides cities and citizens with the means to gather real-time data on air and water quality from thousands of location and to pinpoint problems at the neighborhood or even housing unit level. Efforts such as Air Quality Egg to crowdsource air quality readings and the Floating Sensor Network program at the University of California at Berkeley are showing how low-cost, communicating sensors can be used to gather much more detailed data on what residents are breathing and drinking. Similar approaches can be used to monitor water supply at the tap. Greater awareness and accountability will improve air and water quality. For example, if monitoring leads to effective pollution-reduction strategies, cities could cut particulate matter pollution from 70 micrograms per cubic meter to 20 micrograms per cubic meter. We estimate that this could cut deaths related to air pollution by around 15 percent.

Crime detection and monitoring

Many cities already have security cameras and some have gunshot recognition sensors. IoT will enable these cameras and sensors to automatically detect unusual activities, such as someone leaving a bag unattended, and to trigger a rapid response. Such solutions are already in use in Glasgow, Scotland, and in Memphis, Tennessee, in the United States. Cities that have implemented such systems claim a 10 to 30 percent decrease in crime. We estimate the economic impact of crime reduction on this scale could be more than $30 billion per year.

Transportation

In total, improvements in transportation—measured mostly in time saved by travelers— could have an economic impact of $443 billion to $808 billion per year in 2025.

Centralized and adaptive traffic management

Adaptive traffic control uses real-time data to adjust the timing of traffic lights to improve traffic flow. A centralized control system collects data from sensors installed at intersections to monitor traffic flow. Based on volume, the system adjusts the length of red and green lights to ensure smooth flow. Abu Dhabi recently implemented such a system, which covers 125 main intersections in the city. The system also can give priority to buses, ambulances, or emergency vehicles. For example, if a bus is five minutes behind schedule, traffic signals at the intersection are adjusted to prioritize passage for the bus. Use of adaptive traffic control has been shown to speed traffic flow by between 5 and 25 percent. We estimate that adaptive traffic control and smart meters could reduce time spent in traffic jams and looking for parking spaces by 10 to 15 percent, which could be worth more than $500 billion per year globally in 2025. There could be additional benefits, such as reduced CO2 emissions and postponing or avoiding investment in new roads.

Autonomous vehicles

The use of autonomous vehicles in urban areas can create economic value in a number of ways—freeing up time for drivers; reducing traffic accidents, injuries, and fatalities; saving fuel and raising average highway speeds; and expanding the capacity of parking facilities through self-parking. Autonomous vehicles are already in use in industrial environments such as mines (see worksite setting above). The self-driving passenger car has been in development for several years, and some manufacturers are already offering IoT-based features in production models such as automatic braking for collision avoidance. Some carmakers expect to have self-driving cars on the road by 2020, pending regulatory approval—a non-trivial hurdle. Still, we expect that fully autonomous cars (which require no driver intervention) and partially autonomous cars (which could take over control of all safety-critical functions under certain conditions) to be a reality in cities around the world in 2025. We assume that in 2025, between 1 and 2 percent of light vehicles on the road—

15 million to 30 million vehicles—could be fully self-driving.50 We assume that the penetration of semi-autonomous vehicles could be 12 to 15 percent.

The economic impact of autonomous vehicles in urban settings could be $204 billion to $235 billion per year in 2025. The economic benefit is calculated based on the value of time and fuel saved, traffic fatalities avoided, and greater utilization of assets such as parking spaces. Globally, 1.2 billion people spend 50 minutes on average driving in cars each day. Autonomous vehicles offer the potential to improve traffic flow and free up time spent in the car for other activities. We estimate that time saved through adoption of autonomous vehicles could be worth $15 billion to $25 billion in cities.

In addition, autonomous and partially autonomous vehicles could drastically reduce car accidents. More than 90 percent of US car crashes can be attributed to human error, and more than 40 percent of traffic fatalities involve driver impairment due to alcohol, distraction, drugs, or fatigue. We estimate that traffic accidents could be reduced by 90 percent with the adoption of fully autonomous vehicles and by 40 percent with partially autonomous vehicles, saving 95,000 lives per year, for an estimated economic impact of $180 billion to $200 billion per year.

Autonomous vehicles also can reduce fuel consumption by driving more efficiently. Under computer control, autonomous vehicles would not indulge in wasteful driving behaviors, and with vehicle-to-vehicle communications, cars can travel close together at highway speeds, reducing wind resistance and raising average speed. Autonomous driving could also enable radical changes in automobile design that would make cars lighter and more fuel-efficient. We estimate that fuel consumption could be as much as 15 percent lower.

Finally, because fully autonomous vehicles can park themselves, there is no need to use space between cars in a parking lot or deck to accommodate door openings. This could free up 15 percent of parking space. In addition, autonomous vehicles that drive themselves to parking areas could reduce the need for parking lots and garages in city centers—cars could drop off passengers at their workplaces and even pick up passengers leaving the city center as they proceed to remote parking areas. Adoption of self-driving cars could also lead to new car-pooling and ride-sharing options, which would reduce overall, demand for parking.

Bus and train schedule management

There is a substantial opportunity to save time for riders of public transit by using IoT data. With sensors capturing real-time location data of trains and buses, commuters can shrink the "reliability buffer"—the extra time a traveler builds into a trip to account for possible delays. The buffer can be as much as 70 percent of total trip time.51 Using an app (on a computer or smartphone), commuters can time their exits from home or office (or anywhere in the city) to arrive at the station or bus stop just in time for their trips. In our calculations, we assume that in advanced economies, the average wait time per trip for commuters is 12 minutes; in developing economies, we assume 21 minutes. We further estimate that real-time data could allow commuters to reduce waiting time by approximately 15 percent. Also, given the widespread use of existing GPS-enabled monitoring systems, we assume that development and use of transit-tracking apps will be rapid. Real-time information for buses and trains is already available in New York City, Chicago, Singapore, and some other major cities and is spreading quickly.

We calculate the value of wait time eliminated by looking at the average wage rates in advanced and developing economies and applying a 70 percent discount, which analysts typically use to value non-work time. We arrived at an estimate of $13 billion to $63 billion per year from the potential impact of using IoT to manage bus and train commuting. This does not include other potential gains by transit operators from using IoT data to adjust schedules and routes (reducing service at certain hours or skipping underutilized stops, for example).

There could be additional benefits, such as reduced CO_2 emissions and postponing or avoiding investment in new roads.

Resource/infrastructure management

IoT technology has already demonstrated its potential for monitoring and managing critical urban resources, such as water, sewage, and electric systems. Using sensors to monitor performance across their networks, operators of power and water systems can detect flaws such as leaks in water mains or overheating transformers, enabling operators to prevent costly failures and reduce losses. We estimate that these applications could have an impact of $33 billion to $64 billion per year globally in 2025. Smart meters, which are already being implemented in numerous cities, not only allow utility companies to automate meter reading, but also can enable demand-management programs (encouraging energy conservation through variable pricing, for example), and detect theft of service. By 2025, we estimate that 80 percent of utilities in advanced-economy cities and 50 percent of utilities in developing-economy cities will have adopted smart meters, creating potential value of $14 billion to $25 billion in 2025. Use of IoT technology in distribution and substation automation could have an additional impact of $13 billion to $24 billion per year in 2025. In water systems, we estimate that IoT technology (smart meters) could provide value of $7 billion to $14 billion per year.

Human productivity

The primary ways in which IoT technology would be used to increase productivity of individual workers in urban environments would be through monitoring mobile workers such as motor vehicle operators, building cleaners, pest control workers, and sales representatives. The increased productivity of such workers and IoT-enabled processes to raise productivity of technical and knowledge workers in cities could be worth $2.7 billion to $6.0 billion per year globally in 2025. This assumes an estimated 5 percent increase in productivity of mobile workers and a 3 to 4 percent increase in productivity of knowledge workers.

Enablers and barriers

Multiple factors would need to come together for the Internet of Things to achieve its maximum potential in cities. We expect adoption rates for the IoT applications we size could reach 40 to 80 percent in cities in advanced economies in 2025, and

20 to 40 percent in cities in developing economies. This disparity is a function of having both the ability to fund IoT improvements, which often require public investment, and access to the skills needed for successful implementation and operation of IoT-based systems. The ability to fund IoT investments depends upon the wealth of the city and the government's ability to access investment vehicles or use tax revenue. Another factor is a responsive citizenry. In cities with a high proportion of well-educated residents, demand will likely be higher for the benefits that IoT applications can bring. This can create a virtuous cycle: as successful applications build awareness of benefits, more citizens would demand them.

Achieving maximum potential benefit from IoT in cities also requires interoperability among IoT systems. If autonomous vehicles, a centralized traffic-control system, and smart parking meters were all on speaking terms (so to speak), the commuter's autonomous car could communicate with the centralized traffic system to select the best route, then guide the commuter to the most convenient meter space or the cheapest parking facility where a fully autonomous vehicle could park itself. Interoperability would vastly increase the value of IoT applications in urban settings and encourage many more cities to adopt them.

Cities must also have the technical capacity in their agencies and departments, and committed leadership is essential. To deploy and manage IoT applications requires technical depth that most city governments currently do not possess. Cities that develop this capacity will be ahead in the race to capture IoT benefits. City leaders must also have the political will to drive IoT adoption—to find the funding and make the organizational changes needed to regulate or operate the systems that use IoT technology.

Last but not least, IoT will be broadly adopted only if city governments and the public are assured of the security of IoT-enabled systems. The potential risks are not to be underestimated: malicious parties that find ways to interfere with traffic-control systems or the programs that guide autonomous vehicles could cause enormous damage. Technology vendors will not only need to provide secure systems, but they will also have to convince city governments and residents that the systems truly are secure.

Chapter 7.
IoT Using ARDUINO

Without a doubt, the poster child for the Internet of Things, and physical computing in general, is the Arduino.

These days the Arduino project covers a number of microcontroller boards, but its birth was in Ivrea in Northern Italy in 2005. A group from the Interaction Design Institute Ivrea (IDII) wanted a board for its design students to use to build interactive projects. An assortment of boards was around at that time, but they tended to be expensive, hard to use, or both.

An Arduino Ethernet board, plugged in, wired up to a circuit and ready for use.

So, the team put together a board which was cheap to buy—around £20— and included an onboard serial connection to allow it to be easily pro-grammed. Combined with an extension of the Wiring software environment, it made a huge impact on the world of physical computing.

Wiring: Sketching in Hardware

Another child of the IDII is the Wiring project. In the summer of 2003, Hernando Barragán started a project to make it easier to experiment with electronics and hardware. As the project website (http://wiring.org.co/about. html) puts it:

"The idea is to write a few lines of code, connect a few electronic components to the hardware of choice and observe how a light turns on when person approaches to it, write a few more lines add another sensor and see how this

light changes when the illumination level in a room decreases.

This process is called sketching with hardware—a way to explore lots of ideas very quickly, by selecting the more interesting ones, refining them, and producing prototypes in an iterative process."

The Wiring platform provides an abstraction layer over the hardware, so the users need not worry about the exact way to, say, turn on a GPIO pin, and can focus on the problem they're trying to explore or solve.

That abstraction also enables the platform to run on a variety of hardware boards. There have been a number of Wiring boards since the project started, although they have been eclipsed by the runaway success of the project that took the Wiring platform and targeted a lower-end and cheaper AVR processor: the Arduino project.

A decision early on to make the code and schematics open source meant that the Arduino board could outlive the demise of the IDII and flourish. It also meant that people could adapt and extend the platform to suit their own needs.

As a result, an entire ecosystem of boards, add-ons, and related kits has flourished. The Arduino team's focus on simplicity rather than raw perfor-mance for the code has made the Arduino the board of choice in almost every beginner's physical computing project, and the open source ethos has encouraged the community to share circuit diagrams, parts lists, and source code. It's almost the case that whatever your project idea is, a quick search on Google for it, in combination with the word "Arduino", will throw up at least one project that can help bootstrap what you're trying to achieve. If you prefer learning from a book, we recommend picking up a copy of *Arduino For Dummies*, by John Nussey (Wiley, 2013).

The "standard" Arduino board has gone through a number of iterations: Arduino NG, Diecimila, Duemilanove, and Uno.

The Uno features an ATmega328 microcontroller and a USB socket for connection to a computer. It has 32KB of storage and 2KB of RAM, but don't let those meagre amounts of memory put you off; you can achieve a surprising amount despite the limitations.

The Uno also provides 14 GPIO pins (of which 6 can also provide PWM output) and 6 10-bit resolution ADC pins. The ATmega's serial port is made available through both the IO pins and, via an additional chip, the USB connector.

If you need more space or a greater number of inputs or outputs, look at the Arduino Mega 2560. It marries a more powerful ATmega microcontroller to the same software environment, providing 256KB of Flash storage, 8KB of RAM, three more serial ports, a massive 54 GPIO pins (14 of those also capable of PWM) and 16 ADCs. Alternatively, the more recent Arduino Due has a 32-bit ARM core microcontroller and is the first of the Arduino boards to use this architecture. Its specs are similar to the Mega's, although it ups the RAM to 96KB.

DEVELOPING ON THE ARDUINO

More than just specs, the experience of working with a board may be the most important factor, at least at the prototyping stage. As previously mentioned, the Arduino is optimized for simplicity, and this is evident from the way it is packaged for use. Using a single USB cable, you can not only

The Arduino doesn't, by default, run an OS as such, only the bootloader, which simplifies the code-pushing process described previously. When you switch on the board, it simply runs the code that you have compiled until the board is switched off again (or the code crashes).

It is, however, possible to upload an OS to the Arduino, usually a lightweight real-time operating system (RTOS) such as FreeRTOS/DuinOS. The main advantage of one of these operating systems is their built-in support for multitasking. However, for many purposes, you can achieve reasonable results with a simpler task-dispatching library.

If you dislike the simple life, it is even possible to compile code without using the IDE but by using the toolset for the Arduino's chip—for example, for all the boards until the recent ARM-based Due, the avr-gcc toolset.

The avr-gcc toolset (www.nongnu.org/avr-libc/) is the collection of programs that let you compile code to run on the AVR chips used by the rest of the Arduino boards and flash the resultant executable to the chip. It is used by the Arduino IDE behind the scenes but can be used directly, as well.

Language

The language usually used for Arduino is a slightly modified dialect of C++ derived from the Wiring platform. It includes some libraries used to read and write data from the I/O pins provided on the Arduino and to do some basic handling for "interrupts" (a way of doing multitasking, at a very low level). This variant of C++ tries to be forgiving about the ordering of code; for example, it allows you to call functions before they are defined. This alteration is just a nicety, but it is useful to be able to order things in a way that the code is easy to read and maintain, given that it tends to be written in a single file.

The code needs to provide only two routines:

setup(): This routine is run once when the board first boots. You could use it to set the modes of I/O pins to input or output or to prepare a data structure which will be used throughout the program.

loop(): This routine is run repeatedly in a tight loop while the Arduino is switched on. Typically, you might check some input, do some calculation on it, and perhaps do some output in response.

To avoid getting into the details of programming languages in this chapter, we just compare a simple example across all the boards—blinking a single LED:

```
Pin 13 has an LED connected on most Arduino boards.
give it a name:
int led = 13;
// the setup routine runs once when you press reset: void setup() {

// initialize the         digital pin as an output.
pinMode(led, OUTPUT);
}

// the loop routine        runs over and over again forever:
void loop() {
digitalWrite(led,       HIGH);    // turn the LED on
delay(1000);                      // wait for a second
```

```
digitalWrite(led,        LOW);     // turn the LED off
delay(1000);                        // wait for a second
}
```

Reading through this code, you'll see that the setup() function does very little; it just sets up that pin number 13 is the one we're going to control (because it is wired up to an LED).

Then, in loop(), the LED is turned on and then off, with a delay of a second between each flick of the (electronic) switch. With the way that the Arduino environment works, whenever it reaches the end of one cycle—on; wait a second; off; wait a second—and drops out of the loop() function, it simply calls loop() again to repeat the process.

Debugging

Because C++ is a compiled language, a fair number of errors, such as bad syntax or failure to declare variables, are caught at compilation time. Because this happens on your computer, you have ample opportunity to get detailed and possibly helpful information from the compiler about what the problem is.
Although you need some debugging experience to be able to identify certain compiler errors, others, like this one, are relatively easy to understand:

Blink.cpp: In function 'void loop()':Blink:21: error:'digitalWritee' was not declared in this scope

When the code is pushed to the Arduino, the rules of the game change, however. Because the Arduino isn't generally connected to a screen, it is hard for it to tell you when something goes wrong. Even if the code compiled successfully, certain errors still happen. An error could be raised that can't be handled, such as a division by zero, or trying to access the tenth element of a 9-element list. Or perhaps your program leaks memory and eventually just stops working. Or (and worse) a programming error might make the code continue to work dutifully but give entirely the wrong results.

If Bubblino stops blowing bubbles, how can we distinguish between the following cases?

Nobody has mentioned us on Twitter.
The Twitter search API has stopped working. Bubblino can't connect to the
Internet.
Bubblino has crashed due to a programming error.
Bubblino is working, but the motor of the bubble machine has failed. Bubblino is powered off.

Adrian likes to joke that he can debug many problems by looking at the flashing lights at Bubblino's Ethernet port, which flashes while Bubblino connects to DNS and again when it connects to Twitter's search API, and so on. (He also jokes that we can discount the "programming error" option and that the main reason the motor would fail is that Hakim has poured bubble mix into the wrong hole. Again.) But while this approach might help distinguish two of the preceding cases, it doesn't help with the others and isn't useful if you are releasing the product into a mass market!

The first commercially available version of the WhereDial has a bank of half a dozen LEDs specifically for consumer-level debugging. In the case of an error, the pattern of lights showing may help customers fix their problem or help flesh out details for a support request.

Runtime programming errors may be tricky to trap because although the C++ language has

exception handling, the avr-gcc compiler doesn't support it (probably due to the relatively high memory "cost" of handling exceptions); so the Arduino platform doesn't let you use the usual try...
catch... logic.

Effectively, this means that you need to check your data before using it: if a number might conceivably be zero, check that before trying to divide by it. Test that your indexes are within bounds.

Rear view of a transparent WhereDial. The bank of LEDs can be seen in the middle of the green board, next to the red "error" LED.

But code isn't, in general, created perfect: in the meantime you need ways to identify where the errors are occurring so that you can bullet-proof them for next time. In the absence of a screen, the Arduino allows you to write information over the USB cable using Serial.write(). Although you can use the facility to communicate all kinds of data, debugging information can be particularly useful. The Arduino IDE provides a serial monitor which echoes the data that the Arduino has sent over the USB cable. This could include any textual information, such as logging information, comments, and details about the data that the Arduino is receiving and processing (to double-check that your calculations are doing the right thing).

SOME NOTES ON THE HARDWARE

The Arduino exposes a number of GPIO pins and is usually supplied with "headers" (plastic strips that sit on the pin holes, that provide a convenient solderless connection for wires, especially with a "jumper" connection). The headers are optimised for prototyping and for being able to change the purpose of the Arduino easily.

Each pin is clearly labelled on the controller board. The details of pins vary from the smaller boards such as the Nano, the classic form factor of the Uno, and the larger boards such as the Mega or the Due. In general, you have power outputs such as 5 volts or 3.3 volts (usually labelled 5V and 3V3, or perhaps just 3V), one or more electric ground connections (GND), num-bered digital pins, and numbered analogue pins prefixed with an *A*.

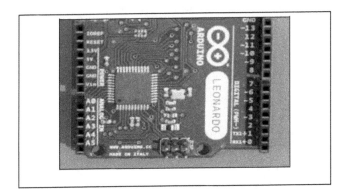

Close-up of an Arduino Leonardo board. Note the labeling of the power and analogue input connections.

You can power the Arduino using a USB connection from your computer. This capability is usually quite convenient during prototyping because you need the serial connection in any case to program the board. The Arduino also has a socket for an external power supply, which you might be more likely to use if you distribute the project. Either way should be capable of powering the microcontroller and the usual electronics that you might attach to it. (In the case of larger items, such as motors, you may have to attach external power and make that available selectively to the component using transistors.)

Outside of the standard boards, a number of them are focused on a particu-lar niche application—for example, the Arduino Ethernet has an on-board Ethernet chip and trades the USB socket for an Ethernet one, making it easier to hook up to the Internet. This is obviously a strong contender for a useful board for Internet of Things projects.

The LilyPad has an entirely different specialism, as it has a flattened form (shaped, as the name suggests, like a flower with the I/O capabilities exposed on its "petals") and is designed to make it easy to wire up with conductive thread, and so a boon for wearable technology projects.

Choosing one of the specialist boards isn't the only way to extend the capabilities of your Arduino. Most of the boards share the same layout of the assorted GPIO, ADC, and power pins, and you are able to piggyback an additional circuit board on top of the Arduino which can contain all manner of componentry to give the Arduino extra capabilities.

In the Arduino world, these add-on boards are called *shields*, perhaps because they cover the actual board as if protecting it.

Some shields provide networking capabilities—Ethernet, WiFi, or Zigbee wireless, for example. Motor shields make it simple to connect motors and servos; there are shields to hook up mobile phone LCD screens; others to provide capacitive sensing; others to play MP3 files or WAV files from an SD card; and all manner of other possibilities—so much so that an entire website, http://shieldlist.org/, is dedicated to comparing and documenting them.

In terms of functionality, a standard Arduino with an Ethernet shield is equivalent to an Arduino Ethernet. However, the latter is thinner (because it has all the components laid out on a single board) but loses the convenient USB connection. (You can still connect to it to push code or communicate over the serial connection by using a supplied adaptor.)

OPENNESS

The Arduino project is completely open hardware and an open hardware success story.

The only part of the project protected is the Arduino trademark, so they can control the quality of any boards calling themselves an Arduino. In addition to the code being available to download freely, the circuit board schematics and even the EAGLE PCB design files are easily found on the Arduino website.

This culture of sharing has borne fruit in many derivative boards being produced by all manner of people. Some are merely minor variations on the main Arduino Uno, but many others introduce new features or form factors that the core Arduino team have overlooked.

In some cases, such as with the wireless-focused Arduino Fio board, what starts as a third-party board (it was originally the Funnel IO) is later adopted as an official Arduino-approved board.

Arduino Case Study: The Good Night Lamp

While at the IDII, Alexandra Deschamps-Sonsino came up with the idea of an Internet-connected table or bedside lamp. A simple, consumer device, this lamp would be paired with another lamp anywhere in the world, allowing it to switch the other lamp on and off, and vice versa. Because light is integrated into our daily routine, seeing when our loved ones turn, for example, their bedside lamp on or off gives us a calm and ambient view onto their lives.

This concept was ahead of its time in 2005, but the project has now been spun into its own company, the Good Night Lamp. The product consists of a "big lamp" which is paired with one or more "little lamps". The big lamp has its own switch and is designed to be used like a normal lamp. The little lamps, however, don't have switches but instead reflect the state of the big lamp.

Adrian was involved since the early stages as Chief Technology Officer. Adrian and the rest of the team's familiarity with Arduino led to it being an obvious choice as the prototyping platform. In addition, as the lamps are designed to be a consumer product rather than a technical product, and are targeted at a mass market, design, cost, and ease of use are also important. The Arduino platform is simple enough that it is possible to reduce costs and size substantially by choosing which components you need in the production version.

A key challenge in creating a mass-market connected device is finding a convenient way for consumers, some of whom are non-technical, to connect the device to the Internet. Even if the user has WiFi installed, entering authentication details for your home network on a device that has no keyboard or screen presents challenges. As well as looking into options for the best solution for this issue, the Good Night Lamp team are also building a version which connects over the mobile phone networks via GSM or 3G. This option fits in with the team's vision of connecting people via a "physical social network", even if they are not otherwise connected to the Internet.

Arduino Case Study: BakerTweet

The BakerTweet device (www.bakertweet.com/) is effectively a physical client for Twitter designed for use in a bakery. A baker may want to let customers know that a certain product has just come out of the ovens—fresh bread, hot muffins, cupcakes laden with icing—yet the environment he would want to tweet from contains hot ovens, flour dust, and sticky dough and batter, all of which would play havoc with the electronics, keyboard, and screen of a computer, tablet, or phone. Staff of design agency Poke in London wanted to know when their local bakery had just produced a fresh batch of their favourite bread and cake, so they designed a proof of concept to make it possible.

Because BakerTweet communicates using WiFi, bakeries, typically not built to accommodate Ethernet cables, can install it. BakerTweet exposes the functionality of Twitter in a "bakery-proof" box with more robust electronics than a general-purpose computer, and a simplified interface that can be used by fingers covered in flour and dough. It was designed with an Arduino, an Ethernet Shield, and a WiFi adapter. As well as the Arduino simply controlling a third-party service (Twitter), it is also hooked up to a custom service which allows the baker to configure the messages to be sent.

Arduino: Getting Started

Arduino is an open-source electronics platform based on easy-to-use hardware and software. Arduino boards are able to read inputs − light on a sensor, a finger on a button, or a Twitter message − and turn it into an output − activating a motor, turning on an LED, publishing something online. You can tell your board what to do by sending a set of instructions to the microcontroller on the board. To do so you use the Arduino programming language (based on Wiring), and the Arduino Software (IDE), based on Processing.

1. Prepare Tools and Parts
You'll need to make sure you have the following items.
Arduino Uno Board
Breadboard – half size
Jumper Wires
USB Cable
LED (5mm)
Push button switch
10k Ohm Resistor
220 Ohm Resistor

2. Download The Software

The Arduino IDE is the interface where we will write the sketches that tell the board what to do.

Download the the latest version of this software on the Arduino, corresponds to your Operating System on desktop.

3. Familiar with IDE

Once the software has been installed on your computer, go ahead and open it up. This is the Arduino IDE and is the place where all the programming will happen.

Menu Bar: Gives you access to the tools needed for creating and saving Arduino sketches.

Verify Button: Compiles your code and checks for errors in spelling or syntax.

Upload Button: Sends the code to the board that's connected such as Arduino Uno in this case. Lights on the board will blink rapidly when uploading.

New Sketch: Opens up a new window containing a blank sketch.

Sketch Name: When the sketch is saved, the name of the sketch is displayed here.
Open Existing Sketch: Allows you to open a saved sketch or one from the stored examples.

Save Sketch: This saves the sketch you currently have open.Serial Monitor: When the board is connected, this will display the serial information of your

ArduinoCode Area: This area is where you compose the code of the sketch that tells the board what to do.

Message Area: This area tells you the status on saving, code compiling, errors and more.

Text Console: Shows the details of an error messages, size of the program that was compiled and additional info.

Board and Serial Port: Tells you what board is being used and what serial port it's connected to.

4. Connect Your Arduino Uno
At this point you are ready to connect your Arduino to your computer. Plug one end of the USB cable to the Arduino Uno and then the other end of the USB to your computer's USB port.Once the board is connected, you will need to go to Tools then Board then finally select Arduino Uno.

Now, you have to tell the Arduino which port you are using on your computer.To select the port, go to Tools then Port then select the port that says Arduino.

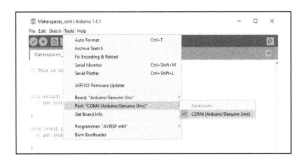

Now, the system to develop the project is ready.

5. Sample Example: Blink an LED

Now, it's time to do our first Arduino project. In this example, we are going to make your Arduino board blink an LED.
Required Parts
Arduino Uno Board R3
Breadboard – half size
Jumper WiresUSB Cable
LED (5mm)
220 Ohm Resistor

Connect The Parts

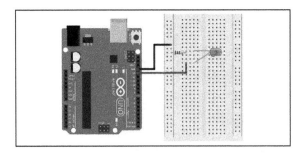

You can build your Arduino circuit by looking at the breadboard image above or by using the written description below. In the written description, we will use a letter/number combo that refs to the location of the component. If we mention H19 for example, that refers to column H, row 19 on the breadboard.

Step 1 – Insert black jumper wire into the GND (Ground) pin on the Arduino and then in the GND rail of the breadboard row 15.

Step 2 – Insert red jumper wire into pin 13 on the Arduino and then the other end into F7 on the breadboard.

Step 3 – Place the LONG leg of the LED into H7.

Step 4 – Place the SHORT leg of the LED into H4.

Step 5 – Bend both legs of a 220 Ohm resistor and place one leg in the GND rail around row 4 and other leg in I4.

Step 6 – Connect the Arduino Uno to your computer via USB cable.
6. Upload The Blink Sketch

Now it's time to upload the sketch (program) to the Arduino and tell it what to do. In the IDE, there are built-in example sketches that you can use which make it easy for beginners.

Arduino is case sensitive.
To open the blink sketch, you will need to go to File > Examples > Basics > Blink

Now you should have a fully coded blink sketch that looks like the image below.

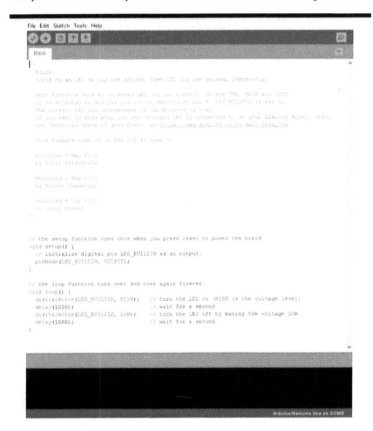

Next, you need to click on the verify button (check mark) that's located in the top left of the IDE box. This will compile the sketch and look for errors.

Once it says "Done Compiling" you are ready to upload it. Click the upload button (forward arrow) to send the program to the Arduino board.

The built-in LEDs on the Arduino board will flash rapidly for a few seconds and then the program will execute. If everything went correctly, the LED on the breadboard should turn on for a second and then off for a second and continue in a loop.

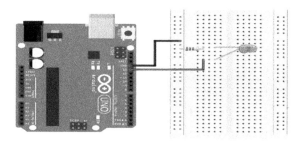

You have just completed your first Arduino project.

Change The Code

Before we go to the next project, lets change some of the code in the "Blink" sketch to make it do something different. Playing around with the sketch will help you start to learn how the code controls the board.

```
// the setup function runs once when you press reset or power the board
void setup() {
  // initialize digital pin LED_BUILTIN as an output.
  pinMode(LED_BUILTIN, OUTPUT);
}

// the loop function runs over and over again forever
void loop() {                          Changed to 200
  digitalWrite(LED_BUILTIN, HIGH);
  delay(200);
  digitalWrite(LED_BUILTIN, LOW);
  delay(200);
}                      Changed to 200
```

Keep the Arduino board connected and change the delay portion of the code from (1000) to (200). Click the verify button on the top left of the IDE and then click upload. This should make the LED on the breadboard blink fast.

Chapter 8.
Implementing IoT Using RASPBERRY PI

The Raspberry Pi, unlike the Arduino, wasn't designed for physical comput-ing at all, but rather, for education. The vision of Eben Upton, trustee and cofounder of the Raspberry Pi Foundation, was to build a computer that was small and inexpensive and designed to be programmed and experimented with, like the ones he'd used as a child, rather than to passively consume games on. The Foundation gathered a group of teachers, programmers, and hardware experts to thrash out these ideas from 2006.

While working at Broadcom, Upton worked on the Broadcom BCM2835 system-on-chip, which featured an exceptionally powerful graphics process-ing unit (GPU), capable of high-definition video and fast graphics rendering. It also featured a low-power, cheap but serviceable 700 MHz ARM CPU, almost tacked on as an afterthought. Upton described the chip as "a GPU with ARM elements grafted on" (www.gamesindustry.biz/articles/digitalfoundry-inside-raspberry-pi).

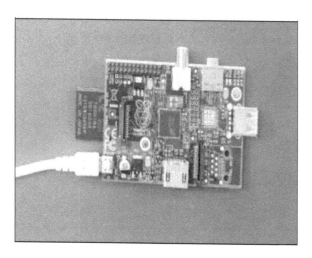

A Raspberry Pi Model B board. The micro USB connector only provides power to the

board; the USB connectivity is provided by the USB host connectors (centre-bottom and

centre-right).

The project has always taken some inspiration from a previous attempt to improve computer literacy in the UK: the "BBC Micro" built by Acorn in the early 1980s. This computer was invented precisely because the BBC producers tasked with creating TV programmes about programming realized that there wasn't a single cheap yet powerful computer platform that was suffi-ciently widespread in UK schools to make it a sensible topic for their show. The model names of the Raspberry Pi, "Model A" and "Model B", hark back to the different versions of the BBC Micro. Many of the other trustees of the Raspberry Pi Foundation, officially founded in 2009, cut their teeth on the BBC Micro. Among them was David Braben, who wrote the seminal game of space exploration, Elite, with its cutting-edge 3D wireframe graphics.

Due in large part to its charitable status, even as a small group, the Founda-tion has been able to deal with large suppliers and push down the costs of the components. The final boards ended up costing around £25 for the more powerful Model B (with built-in Ethernet connection). This is around the same price point as an Arduino, yet the boards are really of entirely different specifications.

The following table compares the specs of the latest, most powerful Arduino model, the Due, with the top-end

	Arduino Due	Raspberry Pi Model B
CPU Speed	84 MHz	700 MHz ARM11
GPU	None	Broadcom Dual-Core VideoCore IV Media Co-Processor
	Arduino Due	**Raspberry Pi Model B**
RAM	96KB	512MB
Storage	512KB	SD card (4GB +)
OS	Bootloader	Various Linux distributions, other operating systems available
Connections	54 GPIO pins 12 PWM outputs 4 UARTs SPI bus I²C bus USB 16U2 + native host 12 analogue inputs (ADC) 2 analogue outputs (DAC)	8 GPIO pins 1 PWM output 1 UART SPI bus with two chip selects I²C bus 2 USB host sockets Ethernet HDMI out Component video and audio out

Raspberry Pi Model B:

So, the Raspberry Pi is effectively a computer that can run a real, modern operating system, communicate with a keyboard and mouse, talk to the Internet, and drive a TV/monitor with high-resolution graphics. The Arduino has a fraction of the raw processing power, memory, and storage required for it to run a modern OS. Importantly, the Pi Model B has built-in Ethernet (as does the Arduino Ethernet, although not the Due) and can also use cheap and convenient USB WiFi dongles, rather than having to use an extension "shield".

Note that although the specifications of the Pi are in general more capable than even the top-of-the-range Arduino Due, we can't judge them as "better" without considering what the devices are for! To see where the Raspberry Pi fits into the Internet of Things ecosystem, we need to look at the process of interacting with it and getting it to do useful physical computing work as an Internet-connected "Thing", just as we did with the Arduino! We look at this next.

However, it is worth mentioning that a whole host of devices is available in the same target market as the Raspberry Pi: the Chumby Hacker Board, the BeagleBoard, and others, which are significantly more expensive. Yes, they may have slightly better specifications, but for the price difference, there may seem to be very few reasons to consider them above the Raspberry Pi. Even so, a project might be swayed by existing hardware, better tool support for a specific chipset, or ease-of-use considerations. In an upcoming section, we look at one such board, the BeagleBone, with regards to these issues.

CASES AND EXTENSION BOARDS

Still, due to the relative excitement in the mainstream UK media, as well as the usual hacker and maker echo chambers, the Raspberry Pi has had some real focus. Several ecosystems have built up around the device. Because the Pi can be useful as a general-purpose computer or media centre without requiring constant prototyping with electronic components, one of the first demands enthusiasts have had was for convenient and attractive cases for it. Many makers blogged about their own attempts and have contributed designs to Thingiverse, Instructables, and others. There have also been several commercial projects. The Foundation has deliberately not authorised an "official" one, to encourage as vibrant an ecosystem as possible, although staffers have blogged about an early, well-designed case created by Paul Beech, the designer of the Raspberry Pi logo (http://shop.pimoroni.com/products/pibow).

Beyond these largely aesthetic projects, extension boards and other accesso-ries are already available for the Raspberry Pi. Obviously, in the early days of the Pi's existence post launch, there are fewer of these than for the Arduino; however, many interesting kits are in development, such as the Gertboard (www.raspberrypi.org/archives/tag/gertboard), designed for conveniently playing with the GPIO pins.

Whereas with the Arduino it often feels as though everything has been done already, in the early days of the Raspberry Pi, the situation is more encourag-ing. A lot of people are doing interesting things with their Pis, but as the platform is so much more high level and capable, the attention may be spread more thinly—from designing cases to porting operating systems to working on media centre plug-ins. Physical computing is just *one* of the aspects that attention may be paid to.

DEVELOPING ON THE RASPBERRY PI

Whereas the Arduino's limitations are in some ways its greatest feature, the number of variables on the Raspberry Pi are much greater, and there is much more of an emphasis on being able to do things in alternative ways. How-ever, "best practices" are certainly developing. Following are some sugges-tions at time of writing. (It's worth checking on the Raspberry Pi websites, IRC channels, and so on, later to see how they will have evolved.)

If you want to seriously explore the Raspberry Pi, you would be well advised to pick up a copy of the Raspberry Pi User Guide, *by Eben Upton and Gareth Halfacree (Wiley, 2012).*

Operating System

Although many operating systems can run on the Pi, we recommend using a popular Linux distribution, such as

Raspbian: Released by the Raspbian Pi Foundation, Raspbian is a distro based on Debian. This is the default "official" distribution and is certainly a good choice for general work with a Pi.
Occidentalis: This is Adafruit's customized Raspbian. Unlike Raspbian, the distribution assumes that you will use it "headless"—not connected to keyboard and monitor—so you can connect to it remotely by default. (Raspbian requires a brief configuration stage first.)

For Internet of Things work, we recommend something such as the Adafruit distro. You're most probably not going to be running the device with a keyboard and display, so you can avoid the inconvenience of sourcing and setting those up in the first place. The main tweaks that interest us are that

The sshd (SSH protocol daemon) is enabled by default, so you can connect to the console remotely.
The device registers itself using zero-configuration networking (zero-conf) with the name raspberrypi.local, so you don't need to know or guess which IP address it picks up from the network in order to make a connection.

When we looked at the Arduino, we saw that perhaps the greatest win was the simplicity of the development environment. In the best case, you simply downloaded the IDE and plugged the device into the computer's USB. (Of course, this elides the odd problem with USB drivers and Internet connec-tion when you are doing Internet of Things work.) With the Raspberry Pi, however, you've already had to make decisions about the distro and down-load it. Now that distro needs to be unpacked on the SD card, which you purchase separately. You should note that some SD cards don't work well with the Pi; apparently, "Class 10" cards work best. The class of the SD card isn't always clear from the packaging, but it is visible on the SD card with the number inside a larger circular "C".

At this point, the Pi may boot up, if you have enough power to it from the USB. Many laptop USB ports aren't powerful enough; so, although the "On" light displays, the device fails to boot. If you're in doubt, a powered USB hub seems to be the best bet.

After you boot up the Pi, you can communicate with it just as you'd commu-nicate with any computer—that is, either with the keyboard and monitor that you've attached, or with the Adafruit distro, via ssh as mentioned previously. The following command, from a Linux or Mac command line, lets you log in to the Pi just as you would log in to a remote server:

```
$ ssh root@raspberrypi.local
```

From Windows, you can use an SSH client such as PuTTY (www.chiark. greenend.org.uk/~sgtatham/putty/). After you connect to the device, you can develop a software application for it as easily as you can for any Linux computer. How easy that turns out to be depends largely on how comfortable you are developing for Linux.

Programming Language

One choice to be made is which programming language and environment you want to use. Here, again, there is some guidance from the Foundation, which suggests Python as a good language for educational programming (and indeed the name "Pi" comes initially from Python).

Let's look at the "Hello World" of physical computing, the ubiquitous "blinking lights" example:

```
import RPi.GPIO as GPIO from time import sleep
GPIO.setmode(GPIO.BOARD) # set the
numbering scheme to be the

#same as on the board
GPIO.setup(8, GPIO.OUT)  #set the GPIO pin 8 to output mode
led = False
PIO.output(8, led)  # initiate the LED to off
while 1: GPIO.output(8, led)
led = not led # toggle the LED status on/off for the next     #iteration
sleep(10)#sleep forone second
```

As you can see, this example looks similar to the C++ code on an Arduino. The only real differences are the details of the modularization: the GPIO code and even the sleep() function have to be specified. However, when you go beyond this level of complexity, using a more expressive "high-level" language like Python will almost certainly make the following tasks easier:

Handling strings of character data

Completely avoiding having to handle memory management (and bugs related to it) Making calls to Internet services and parsing the data received Connecting to databases and more complex processing

Abstracting common patterns or complex behaviours

Also, being able to take advantage of readily available libraries on PyPi (https://pypi.python.org/pypi) may well allow simple reuse of code that other people have written, used, and thoroughly tested.

So, what's the catch? As always, you have to be aware of a few trade-offs, related either to the Linux platform itself or to the use of a high-level programming language. Later, where we mention "Python", the same considerations apply to most higher-level languages, from Python's contem-poraries Perl and Ruby, to the compiled VM languages such as Java and C#. We specifically contrast Python with C++, as the low-level language used for Arduino programming.

Python, as with most high-level languages, compiles to relatively large (in terms of memory usage) and slow code, compared to C++.

The former is unlikely to be an issue; the Pi has more than enough memory. The speed of execution may or may not be a problem: Python is likely to be "fast enough" for most tasks, and certainly for anything that involves talking to the Internet, the time taken to communicate over the network is the major slowdown. However, if the electronics of the sensors and actuators you are working with require split-second timing, Python might be too slow. This is by no means certain; if Bubblino starts blowing bubbles a millisecond later, or the DoorBot unlocks the office a millisecond after you scan your RFID card to authenticate, this delay may be acceptable and not even noticeable.

Python handles memory management automatically. Because handling the precise details of memory allocation is notoriously fiddly, automatic memory management generally results in fewer bugs and performs adequately. However, this automatic work has to be scheduled in and takes some time to complete. Depending on the strategy for garbage collection, this may result in pauses in operation which might affect timing of subsequent events.

Also, because the programmer isn't exposed to the gory details, there may well be cases in which Python quite reasonably holds onto more memory than you might have preferred had you been managing it by hand. In worse cases, the memory may never be released until the process terminates: this is a so-called memory leak. Because an Internet of Things device generally runs unattended for long periods of time, these leaks may build up and eventually end up with the device running out of memory and crashing. (In reality, it's more likely that such memory leaks happen as a result of programming error in manual memory management.)

Linux itself arguably has some issues for "real-time" use. Due to its being a relatively large operating system, with many processes that may run simultaneously, precise timings may vary due to how much CPU priority is given to the Python runtime at any given moment. This hasn't stopped many embedded programmers from moving to Linux, but it may be a consideration for your case.
An Arduino runs only the one set of instructions, in a tight loop, until it is turned off or crashes. The Pi constantly runs a number of processes. If one of these processes misbehaves, or two of them clash over resources (memory, CPU, access to a file or to a network port), they may cause problems that are entirely unrelated to your code. This is unlikely (many well-run Linux computers run without maintenance for years and run businesses as well as large parts of the Internet) but may result in occasional, possibly intermittent, issues which are hard to identify and debug.

We certainly don't want to put undue stress on the preceding issues! They are simply trade-offs that may or may not be important to you, or rather more or less important than the features of the Pi and the access to a high-level programming language.

The most important issue, again, is probably the ease of use of the environ-ment. If you're comfortable with Linux, developing for a Pi is relatively simple. But it doesn't approach the simplicity of the Arduino IDE. For example, the Arduino starts your code the moment you switch it on. To get the same behaviour under Linux, you could use a number of mechanisms, such as an initialization script in /etc/init.d.

First, you would create a wrapper script—for example, /etc/init.d/ StartMyPythonCode. This script would start your code if it's called with a start argument, and stop it if called with stop. Then, you need to use the chmod command to mark the script as something the system can run: chmod +x /etc/init.d/StartMyPythonCode. Finally, you register it to run when the machine is turned on by calling sudo update-rc.d StartMyPythonCode defaults.

If you are familiar with Linux, you may be familiar with this mechanism for automatically starting services (or indeed have a preferred alternative). If not, you can find tutorials by Googling for "Raspberry Pi start program on boot" or similar. Either way, although setting it up isn't hard per se, it's much more involved than the Arduino way, if you aren't already working in the IT field.

Debugging

While Python's compiler also catches a number of syntax errors and attempts to use undeclared variables, it is also a relatively permissive language (compared to C++) which performs a greater number of calculations at runtime. This means that additional classes of programming errors won't cause failure at compilation but will crash the program when it's running, perhaps days or months later.

Whereas the Arduino had fairly limited debugging capabilities, mostly involving outputting data via the serial port or using side effects like blinking lights, Python code on Linux gives you the advantages of both the language and the OS. You could step through the code using Python's integrated debugger, attach to the process using the Linux strace command, view logs, see how much memory is being used, and so on. As long as the device itself hasn't crashed, you may be able to ssh into the Raspberry Pi and do some of this debugging while your program has failed (or is running but doing the wrong thing).

Because the Pi is a general-purpose computer, without the strict memory limitations of the Arduino, you can simply use try... catch... logic so that you can trap errors in your Python code and determine what to do with them. For example, you would typically take the opportunity to log details of the error (to help the debugging process) and see if the unexpected problem can be dealt with so that you can continue running the code. In the worst case, you might simply stop the script running and have it restart again afresh!

Python and other high-level languages also have mature testing tools which allow you to assert expected behaviours of your routines and test that they perform correctly. This kind of automated testing is useful when you're working out whether you've finished writing correct code, and also can be rerun after making other changes, to make sure that a fix in one part of the code hasn't caused a problem in another part that was working before.

SOME NOTES ON THE HARDWARE

The Raspberry Pi has 8 GPIO pins, which are exposed along with power and other interfaces in a 2-by-13 block of male header pins. Unlike those in the Arduino, the pins in the Raspberry Pi aren't individually labelled. This makes sense due to the greater number of components on the Pi and also because the expectation is that fewer people will use the GPIO pins and you are discouraged from soldering directly onto the board. The intention is rather that you will plug a cable (IDC or similar) onto the whole block, which leads to a "breakout board" where you do actual work with the GPIO.

Alternatively, you can connect individual pins using a female jumper lead onto a breadboard. The pins are documented on the schematics. A female-to-male would be easiest to connect from the "male" pin to the "female" breadboard. If you can find only female-to-female jumpers, you can simply place a header pin on the breadboard or make your own female-to-male jumper by connecting a male-to-male with a female-to-male! These jumpers are available from hobbyist suppliers such as Adafruit, Sparkfun, and Oomlout, as well as the larger component vendors such as Farnell.

The block of pins provides both 5V and 3.3V outputs. However, the GPIO pins themselves are only 3.3V tolerant. The Pi doesn't have any over-voltage protection, so you are at risk of breaking the board if you supply a 5V input! The alternatives are to either proceed with caution or to use an external breakout board that has this kind of protection. At the time of writing, we can't recommend any specific such board, although the Gertboard, which is mentioned on the official site, looks promising.
Note that the Raspberry Pi doesn't have any analogue inputs (ADC), which means that options to connect it to electronic sensors are limited, out of the box, to digital inputs (that is, on/off inputs such as buttons). To get readings from light-sensitive photocells, temperature sensors, potentiometers, and so on, you need to connect it to an external ADC via the SPI bus. You can find instructions on how to do this at, for example, http://learn.adafruit.com/reading-a-analog-in-and-controlling-audio-volume-with-the-raspberry-pi/overview.
We mentioned some frustrations with powering the Pi earlier: although it is powered by a standard USB cable, the voltage transmitted over USB from a laptop computer, a powered USB hub, or a USB charger varies greatly. If you're not able to power or to boot your Pi, check the power requirements and try another power source.

OPENNESS

Because one of the goals of the Raspberry Pi is to create something "hack-able", it is no surprise that many of the components are indeed highly open: the customized Linux distributions such as "Raspbian" (based on Debian), the ARM VideoCore drivers, and so on. The core Broadcom chip itself is a proprietary piece of hardware, and they have released only a partial data-sheet for the BCM2835 chipset. However, many of the Raspberry Pi core team are Broadcom employees and have been active in creating drivers and the like for it, which are themselves open source.
These team members have been able to publish certain materials, such as PDFs of the Raspberry Pi board schematics, and so on. However, the answer to the question "Is it open hardware?" is currently "Not yet" (www.raspberrypi. org/archives/1090#comment-20585).

The Raspberry Pi is a pocket-friendly and pocket-sized version of a Central Processing Unit, complete with a processor, memory and RAM among other components. Originally developed in the UK, the rpi was intended to act as a means to provide processing power to those individuals for whom the traditional computers were way too costly or just out of reach. It, thus, opened up a new gateway of possibilities to the common user, thereby creating a revolution of its own in the field of technology.

Implementation of IoT using Raspberry Pi

The Raspberry Pi is a pocket-friendly and pocket-sized version of a Central Processing Unit, complete with a processor, memory and RAM among other components. Originally developed in the UK, the rpi was intended to act as a means to provide processing power to those individuals for whom the traditional computers were way too costly or just out of reach. It, thus, opened up a new gateway of possibilities to the common user, thereby creating a revolution of its own in the field of technology.

Recent innovations of various RPi models have led to a promising start of an affordable yet powerful, automated system that can be applied in multiple domains. Showing some serious potential for smart-systems, the RPi may hold the key to creating efficient, autonomous computing sub-systems.

The capabilities of this wonderful device, however, are not restricted to automation only. The GPIO (General Purpose Input/output) in particular, is a very crucial part of the RPi. It allows the attaining of useful information through a logical data-flow mechanism and processes it to decide the RPi's response. It consists of various pins that provide a means to connect the necessary hardware with the RPi. This paper presents a basic application of RPi in the field of home automation and Internet of Things, in which the control signal of the respective GPIO pin of the Raspberry Pi is controlled by the status of the appliance stored in a MySQLite Database table as binary values which is read by the Python script that is run on the RPi continuously as soon as it boots. The interfaces used to change the status of the appliance in the database are a custom-made Android App or a web-page.

This status value is stored in the database as 0 or 1.Depending on this value, the appliance is either turned ON or OFF. The purpose of the python script, here, is to manipulate the signal via GPIO and set it as either high or low based on user-action through the Android

App or Web-browser. A multimeter is used to effectively check the actual status (ON/OFF) of the GPIO pins in the RPi and thus, confirm that the device is working successfully.

Getting started with the Windows 10 IoT Core on Raspberry Pi

v

The Raspberry Pi is a small computer that can do lots of things. We can plug it into a monitor and attach a keyboard and mouse.

March 14 is known as Pi Day because the date represents the first three numbers in the mathematical constant π (3.14). We're celebrating with our coverage of everything Raspberry Pi related. If you've never even thought of what HTML means, you can still create amazing gadgets using Raspberry Pi and a bit of imagination.

Step 1: Get the Board and Tools Ready
The things required are-

1. A PC running Windows 10

2. Raspberry Pi Model A/B/B+

3. Raspberry Pi 5v Micro USB power supply with at least 1.0A current. If you plan on using several power-hungry USB peripherals, use a higPher current power supply instead (>2.0A).

4. 8GB Micro SD card – class 10 or better.

5. HDMI cable and monitorEthernet

6. CableMicroSD card reader – due to an issue with most internal micro SD card readers, we suggest an external USB micro SD card reader.

Raspberry pi- A/B/B+

Ethernet cable   Usb power cord   86b Sd card with Adapter

Step 2: Install the Windows 10 IoT Core Tools

1. Download a Windows 10 IoT Core image from this <u>Download</u> page. Save the ISO to a local folder on your desktop/laptop.
2. Double click on the ISO (Iot Core RPi.iso). It will automatically mount itself as a virtual drive so you can access the contents.

3. Install Windows_10_IoT_Core_RPi2.msi. When installation is complete, flash.ffu will be located at C:\Program Files (x86)\Microsoft IoT\FFU\RaspberryPi2.

4. Eject the Virtual CD when installation is complete – this can be done by navigating to the top folder of File Explorer, right clicking on the virtual drive, and selecting "Eject".

Step 3: Put the Windows 10 IoT Core Image on Your SD Card
1. Insert a Micro SD Card into your SD card reader.

2. Use IoTCoreImageHelper.exe to flash the SD card. Search for "Windows IoT" from start menu and select the shortcut "WindowsIoTImageHelper".

3. The tool will enumerate devices as shown. Select the SD card you want to flash, and then provide the location of the ffu to flash the image.

NOTE: IoTCoreImageHelper.exe is the recommended tool to flash the SD card.

4. Safely remove your USB SD card reader by clicking on "Safely Remove Hardware" in your task tray, or by finding the USB device in File Explorer, right clicking, and choosing "Eject". Failing to do this can cause corruption of the image.

Step 4: Plug in Board

1. Insert the micro SD card you prepared into your Raspberry Pi 2 (the slot is indicated by arrow #1 in the image below).

2. Connect a network cable from your local network to the Ethernet port on the board.

3. Make sure your development PC is on the same network.

4. Connect an HDMI monitor to the HDMI port on the board.

5. Connect the power supply to the micro USB port on the board.

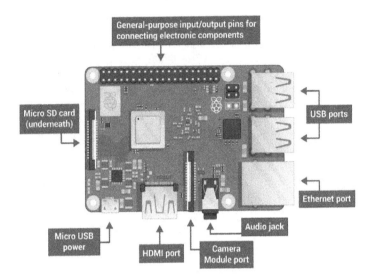

General-purpose input/output pins for connecting electronic components

Micro SD card (underneath)

USB ports

Ethernet port

Audio jack

Micro USB power

HDMI port

Camera Module port

USB ports – these are used to connect a mouse and keyboard. You can also connect other components, such as a USB drive.
SD card slot – you can slot the SD card in here. This is where the operating system software and your files are stored.
Ethernet port – this is used to connect the Raspberry Pi to a network with a cable. The Raspberry Pi can also connect to a network via wireless LAN.
Audio jack – you can connect headphones or speakers here.
HDMI port – this is where you connect the monitor (or projector) that you are using to display the output from the Raspberry Pi. If your monitor has speakers, you can also use them to hear sound.
Micro USB power connector – this is where you connect a power supply. You should always do this last, after you have connected all your other components.
GPIO ports – these allow you to connect electronic components such as LEDs and buttons to the Raspberry Pi.

Step 5: Boot Windows 10 IoT Core
1. Windows 10 IoT Core will boot automatically after connecting the power supply. This process will take a few minutes. After seeing the Windows logo, your screen may go black for about a minute – don't worry, this is normal for boot up. You may also see a screen prompting you to choose a language for your Windows 10 IoT Core device – either connect a mouse and choose your option, or wait about a 1-2 minutes for the screen to disappear.

2. Once the device has booted, the DefaultApp will launch and display the IP address of your RPi2.

Step 6: Connecting to Your Device

1.You can use Windows Device Portal to connect to your device through your favorite web browser. The device portal provides configuration and device management capabilities, in addition to advanced diagnostic tools to help you troubleshoot and view the real time performance of your Windows IoT Device.

NOTE: It is highly recommended that you update the default password for the Administrator account.

Step 7: Finish Off

Now you should be ready with a windows 10 raspberry pi.

Browsing the web

You might want to connect your Raspberry Pi to the internet. If you didn't plug in an ethernet cable or connect to a WiFi network during the setup, then you can connect now.

Click the icon with red crosses in the top right-hand corner of the screen, and select your network from the drop-down menu. You may need to ask an adult which network you should choose.

- Type in the password for your wireless network, or ask an adult to type it for you, then click **OK**.

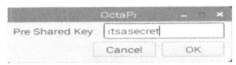

- When your Pi is connected to the internet, you will see a wireless LAN symbol instead of the red crosses.

- Click the web browser icon and search for raspberry pi.

When your Pi is connected to the internet, you will see a wireless LAN symbol instead of the red crosses.

Click the web browser icon and search for raspberry pi.

Creating a Raspberry PI app with Visual Studio on Windows 10 IoT Core

This article demonstrates how to build and debug a simple Raspberry PI application using Visual Studio. Earlier we have seen the seen and install Windows 10 IoT Core on Raspberry pi 3.

1. Download (http://gnutoolchains.com/raspberry) Windows toolchain for Raspberry PI and install it by running the installer on your desktop (Windows 10).

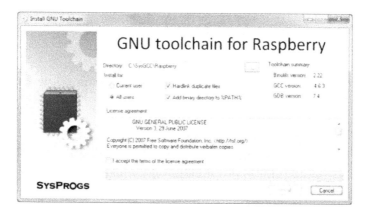

2. If you have not prepared the SD card yet, download and unpack (http://downloads.raspberrypi.org/raspbian_latest) the SD card image.

3. Go to the directory where you have installed the toolchain and run **tools\WinFLASHTool.exe**

4. Specify the path to the SD card image you have downloaded and unpacked.

5. Insert your SD card into the card reader and select it in the list:

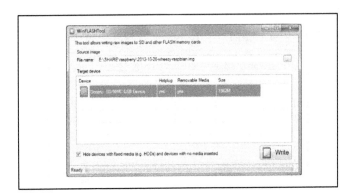

6. Press the Write button. Double-check the information in the confirmation dialog to avoid confusing the SD card with another storage device and rewriting it by mistake:

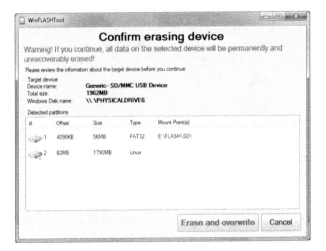

7. Wait until WinFLASHTool finishes writing the image to your SD card. Writing the SD card typically takes several minutes depending on the card speed:

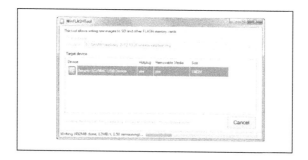

8. Insert the SD card into the slot on the Raspberry PI board. Connect the network cable to connect it with desktop. Optionally connect the keyboard and the HDMI cable. Finally connect the power cable and let the board start:

9. Use the keyboard/screen (ifconfig command) or the DHCP status viewer of your router to find out the IP address of the Raspberry PI board. In this tutorial it is **192.168.0.114**

10. Please download (https://visualgdb.com/download)and install the latest VisualGDB on your desktop which is compatible with **Visual Studio 2008-19.**

10.   On your Windows machine start Visual Studio, select "File->New project". Then select "VisualGDB->Linux Project Wizard". Specify project location and press "OK".

12. The VisualGDB Linux Project Wizard will start. As we are making a simple "Hello, World" application, keep "Create a new project" selected and press "Next".

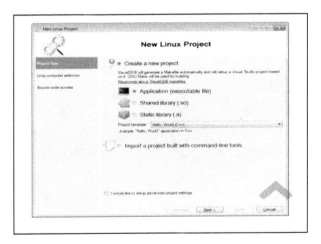

13. If you have not created any Raspberry PI projects before, select "Create a new SSH connection" on the next page.

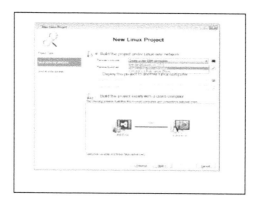

14. Provide the IP address of your Raspberry PI box, the user name ("pi" by default) and password ("raspberry" by default). It is recommended to check the "setup public key" checkbox, so that VisualGDB will automatically generate an public/private keypair, store it in your Windows account's key container and setup the Raspberry PI box to use it.

If you don't enable public key authentication, VisualGDB will remember your password for this connection. The stored passwords are encrypted using a key stored in your Windows account. Thus, the password will only be available once you login using your Windows account.

15. You could use 2 options to build your first Raspberry PI app: build it on Windows with a cross-compiler or build it on the Raspberry PI itself. The first option is faster, while the second is easier to setup. In this tutorial we will use the second option:

Use the diagram at the bottom of the page to check the correctness of your setup:
a. The hammer icon corresponds to the machine where the compiler is run.
b. The "play" icon corresponds to the machine where the debugged program is launched.

16. When you press "Next", VisualGDB will test your toolchain by trying to compile and run a trivial program. If any errors are detected at this stage, you will see a detailed error log with further troubleshooting information.

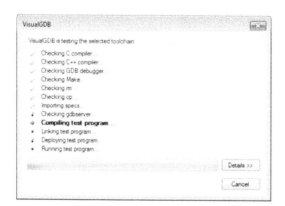

17. As the same source code will be edited under Windows and then compiled under Linux, VisualGDB will need to keep the sources synchronized. The easiest way is to use automatic file uploading via SSH.

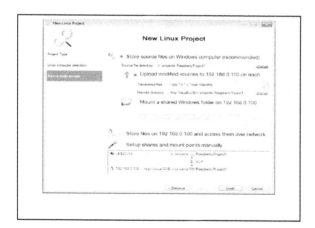

18. Press Finish to complete the wizard. If this is your first project for Raspberry PI, VisualGDB will cache the include directories to make them available through IntelliSense.

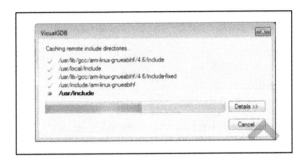

19. Replace the contents of the main source file with the following:

```
1  #include <stdio.h>
2  #include <unistd.h>
3
4  int main(int argc, char *a
5  {
6      char szHost[128];
7      gethostname(szHost, si
8      printf("The host name
9      return 0;
10 }
```

20. Build the solution. You will see how the source files are transferred to Raspberry PI and compiled there:

21. Set a breakpoint on the printf() line and press F5 to start debugging. The breakpoint will trigger:

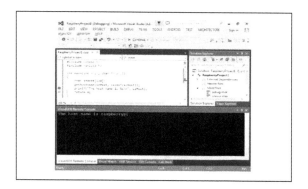

22. VisualGDB provides an advanced Clang-based IntelliSense engine that is fully compatible with GNU language extensions and provides advanced features like refactoring, Code Map and auto completion in the Watch windows:

23. You can customize various project settings by right-clicking on the project in Solution Explorer and selecting VisualGDB Project Properties:

You can now use all normal Visual Studio techniques to debug your app.

SMART PREMISES

Internet of Things

The Internet of Things (IoT) , in its most basic form, can be described as a system comprised of various "things" that are interconnected to each other for the sole purpose of creating, sending, manipulating and processing data.

A thing, in this definition, could be any device, machine, object, or an organism capable of doing one or all of the aforementioned tasks. Although it was perceived as an idea quite a long time ago, the IoT did not officially come into the picture of technological advancements until the 90's which was also around the same time that there was a major boom in the devices using Internet and the web itself.

With the popularity of the Internet and the public welcoming various new horizons of the modern world into their daily life, it became possible for both commercial and non-commercial developers and innovators to create devices and systems that were previously not explored, either due to being unpopular or having limited functionality. But now, every device that was on the Internet was in one way or another, part of an IoT system without even realizing it. The model of the World Wide Web was actually synonymous with a spider's web and it could be said that each device that was connected, was passing information on the web that could be used by another device on the web to perform certain tasks. But, this group of connected devices would need to be logical, in order to work as an efficient system.

For example, integrating the temperature sensor in your room to your AC system would be a nice touch of automation. But, trying to integrate the garage door system with your bathtub would make no sense at all. The key here is trying to understand, which of the systems or sub-systems would benefit through integration and which ones are better left alone. Once we are able to identify this, we regulate the data flow from one-device to another and hence, create a chain of linked devices, that can not only function as individual entities, but also work as a module of a group to generate new data and perform operations based on input from other similar devices. In essence, we create an IoT system that can work autonomously or have some extent of control by waiting for user-action to do something.

Premise Automation

The concept of a fully automated, smart premise is one of the most popular technologies out there in the world which has been praised by both the Common User and Tech Gurus alike. The sudden increase in the development and innovation of such systems/sub-systems, of course, is due to the extent of control and ease we have gained from the Internet.

With the Internet growing into an ever expanding Universe of more and more people joining it per second, we have also started using the Internet for more than just a Google search or posting our favorite moments for others to see. The true potential of the Internet lies in the amount of data that it can generate and support in terms of the processing power and storage facilities among others. Connecting and controlling multiple devices by adding them to a common interface that flows through all of them seamlessly, is pretty much the definition of the IoT and when combined with multiple platforms like Web and Android, we can further stretch the limits of what such a system can achieve through a simple User-Action. This connection and control is not confined to a few devices, however. With the right kind of tools, we can grow this feature by moving into premise automation.

An automated/smart premise (like a Smart Home or Smart Hotel) is pretty much a physical area consisting of multiple electronic-devices that we use in our daily lives. No matter what the size or purpose of the device is, if it is on the grid and connected to its peers, you can control it remotely with a few simple actions ranging from a simple hand gesture to a complete biometric authentication.

The most uncomplicated example of this, though, is the one described below, where a User can control multiple devices using one touch or click.

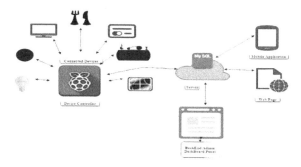

Software Implementation

The RPi runs on Raspbian O.S. which is based on the Debian Wheezly Linux O.S. and uses a Python script to operate the appliances connected to it. The user has the following two interface options to send device data to the database in real-time which is read by the Python script continuously on the RPi :

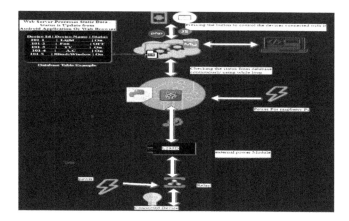

Web Page

For development of the web-page implemented on the user-end, HTML5 & JavaScript were used. On the server-end we have PHP to read static data containing information about the GPIO ports that are in use and the devices connected to each room in JSON format. The web-page offers control of remote devices via three main interfaces which are as follows :

Login Form : The login form is a simple PHP script that interacts with a MySQLite database. On successful login, the user can see a screen that lets him/her select a room.

Screen For Selecting A Room : On this screen, the user selects a room of their choice from the entire premise layout available. As soon as the user does this, he/she will get a pop-up to enter credentials provided to them at the time of registration (like hotel check-ins etc). The following information is stored about the rooms in JSON format:
> **Room Name/Room No**
> **Description**
> **Type**

Each room is represented by a JSON object containing the above mentioned data for each room.

Devices : This screen shows the information about the room selected in the previous screen and all the connected devices in the room. Each individual device is represented by an icon that displays the type, name, description and status of that device. Other than these, there will be two separate operational buttons attached to each device. These two buttons will be labeled ON and OFF and used to remotely turn a particular device ON or OFF.

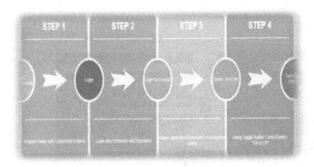

Android Application

The working of the Android Application is similar in function to the web-page to some extent. Initially, the user has to login to the app using a login screen. After the successful login of user, he/she will be asked for an extra credential provided at the time of registration. After this authentication, the user will be able to see the selected room of their choice. The room will further contain icons/images representing various devices that are interconnected to each other inside the room. The device that is selected by user will display certain information like type, name, description and status of the device. The device status can be manipulated by using one of the two toggle buttons labeled ON and OFF respectively for each individual device. The user will then toggle one of the two buttons to send status value for that device to the MySQLite database via the PHP script.The data gathered from the User action through any of the two interfaces (Web , or Android) will ultimately be passed on to the MySQLite database and later read by the Python script running continuously on the

RPi to actually switch ON or OFF the device.

Hardware Implementation

The integration of electronic components into the Raspberry-pi we have used are: Raspberry Pi 3 Model B, power supply module for Raspberry Pi, 830 Tie Points MB102 Breadboard, Bulb, 8 Channel DC 5V Relay Module for RPi, Smartphone, a internet connection for both Smartphone and Raspberry Pi.

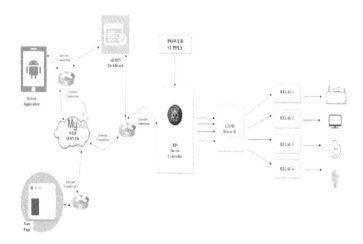

OTHER APPLICATIONS

There some more really interesting areas where the RPi can be implemented. Following is a list of such possible applications of RPi :

Remote Camera: When connected to a camera, the RPi has the ability to transmit images to end-user remotely by responding to an event or command initiated by the user.

Chat Bot: The ability of the RPi to accept and respond to content from the end user makes it an ideal candidate for its implementation as a bot. To make things further interesting, you can also create another bot instead of an actual user sending data to the RPi.

Digital Advertisement Screen: The RPi can also be used to display digitally created advertisements like video/animation etc by simply connecting it to a display device or a projector. Any change in display screen we can edit or operate remotely by owner/user from anywhere, anytime using internet.

Android TV: Another interesting application of the RPi is that you can turn your regular TV into an Android TV by simply integrating it with an RPi having an Android TV operating system.

Automated Printer Machine: One of the most innovative ideas is to create a smart photo-copy/printing machine that accepts all parameters like number of pages, print format etc from the user, along with a payment for the photo-copy/print and automatically prints the required number of pages for the customer. It would even display messages like out of ink/paper etc. Also we can add credentials (like Barcode Reader, RFID card) to the machine for authenticating a free user or employee in the system.

Customizable Device: The RPi can be customized to create multi-purpose systems like a visual device such as a Tablet and personalized audio-video systems that include a combination of various display screens and audio-cards. Also, the RPi can be tweaked to create a streaming device for your network requirements with a highly-configurable design.

Physical Computing Device: The RPi can serve the purpose of a physical computing device that interacts with the outside world using many digital/virtual projects like home media centers, arcade gaming centers or custom-made servers

REFERENCES:

1) Benisch, M., Kelley, P. G., Sadeh, N. & Cranor, L. F. Capturing location privacy preferences: quantifying accuracy and user-burden tradeofs. Personal Computing, in 15, (2011), 679-694.

2) Patel, S. N., Gupta, S. & Reynolds, M. S. The design and evaluation of an end-user-deployable, whole house, contactless power consumption sensor. In Proc. CHI 2010, ACM Press (2010), 2471-2480.

3) Gaver, W. W., Bowers, J., Boehner, K., Boucher, A., Cameron, D. W. T, Hauenstein, M., Jarvis, N. & Pennington, S. Indoor weather stations: investigating a ludic approach to environmental HCI through batch prototyping. In Proc. CHI 2013, ACM Press (2013), 3451-3460.

4) Grosse-Puppendahl, T., Braun, A., Kamieth, F. & Kuijper, A. Swiss-Cheese extended: an object recognition method for ubiquitous interfaces based on capacitive proximity sensing. In Proc. CHI 2013, ACM Press (2013), 1401-1410.

5) Harrison, C., Tan, D. & Morris, D. Skinput: appropriating the body as an input surface. In Proc. CHI 2010, ACM Press (2010), 453-462.

6) Brush, A. J. B., Lee, B., Mahajan, R., Agarwal, S., Saroiu, S. & Dixon, C. Home automation in wild: challenges and opportunities. In Proc. CHI 2011, ACM Press (2011), 2115-2125.

7) Botterman, M. Internet of Things: an early reality of the Future Internet. Report of Internet of Things workshop. European Commission Information Society and Media Directorate General (2009) at Pragua.

8) Carretero, J. & García, J. D. The Internet of Things: connecting the world. Personal Ubiquitous Computing (2013).

9) Costanza, E., Panchard, J., Zufferey, G., Nembrini, J. Freudiger, J., Huang, J. & Hubaux, J. SensorTune: mobile

10) http://www.raspberrypi.org/.

11) Jaosn Lengstorf, Phil Leggetter, Book: "Realtime Web Apps with HTML5 WebSocket, PHP, and jQuery", 2013.

12) http://tools.ietf.org/html/rfc6455.

13) G. Kortuem, F. Kawsar, D. Fitton, and V. Sundramoorthy, "Smart objects as building blocks for the internet of things," Internet Computing, IEEE, vol. 14, pp. 44-51, 2010.

14) http://whatis.techtarget.com/definition/Internet-of-Things

15) http://en.wikipedia.org/wiki/Home_automation.

16) Soundhar Ganesh, Venkatash, Vidhyasagar, Maragatharaj,

17) "Raspberry Pi Based Interactive Home Automation System through Internet of

Things", International Journal for Research in Applied Science & Engineering Technology (IJRASET), Vol: 3, Issue 3, March - 2015, pp: 809-814

18) http://developer.android.com.

19) http://pi4j.com/

20) DIY Hacking, "IOT Based Raspberry Pi Home Automation using IBM Bluemix", http://diyhacking.com /raspberry-pihome-automation-ibm-bluemix/.

21) Dr. S. Kanaga Suba Raja, C. Viswatnathan, Dr. D.
22) Sivakumar, M. Vivekanandan, "Secured Smart Home Energy System (SSHEMS) Using Raspberry Pi", Journal of
23) Theoretical and Applied Information Technology, 10 th August 2014. Vol: 66 No.1, pp: 305-314.

[IoT] Ovidiu Vermesan & Peter Fress, "Internet of Things –From Research and Innovation to Market Deployment", River Publishers Series in Communication, ISBN: 87-93102-94-1, 2014.

[SusAgri] Rodriguez de la Concepcion, A.; Stefanelli, R.; Trinchero, D. "Adaptive wireless sensor networks for high-definition monitoring in sustainable agriculture", Wireless Sensors and Sensor Networks (WiSNet), 2014

[sixlo] IETF WG 6Lo: http://datatracker.ietf.org/wg/6lo/charter/

[sixlowpan] IETF WG 6LoWPAN: http://datatracker.ietf.org/wg/6lowpan/charter/

[IANA-IPV6-SPEC] IANA IPv6 Special-Purpose Address Registry: http://www.iana.org/ assignments/iana-ipv6-special-registry/

[IEEE802.15.4] IEEE Computer Society, "IEEE Std. 802.15.4-2003", October 2003

[RFC2460] S. Deering, R. Hinden, "Internet Protocol, Version 6 (IPv6) Specification", December 1998, RFC 2460, Draft Standard

[RFC3315] Droms, R., Bound, J., Volz, B., Lemon, T., Perkins, C., and M. Carney, "Dynamic Host Configuration Protocol for IPv6 (DHCPv6)", July 2003

[RFC4193] R. Hinden, B. Haberman, "Unique Local IPv6 Unicast Addresses", RFC4193, October 2005

[RFC4291] R. Hinden, S. Deering, "IP Version 6 Addressing Architecture", RFC 4291, February 2006

[RFC4862] S. Thomson, T. Narten, T. Jinmei, "IPv6 Stateless Address Autoconfiguration", September 2007

[RFC6775] Z. Shelby, Ed., S. Chakrabarti, E. Nordmárk, C. Bormann, "Neighbor Discovery Optimization for IPv6 over Low-Power Wireless Personal Area Networks (6LoWPANs)", November 2012, RFC 6775, Proposed Standard

[RFC6890] M. Cotton, L. Vegoda, R. Bonica, Ed., B. Haberman, "Special-Purpose IP Address Registries", RFC 6890 / BCP 153, April 2013

[roll] roll IETF WG: http://datatracker.ietf.org/wg/roll/charter

[I802154r] L. Frenzel, "What's The Difference Between IEEE 802.15.4 And ZigBee Wireless?" [Online] Available: http://electronicdesign.com/what-s-difference-between/what-s-difference-between-ieee-802154-and-zigbee-wireless.

BIBLIOGRAPHY:

A

ABC News, "100 million dieters, $20 billion: The weight-loss industry by the numbers," May 8, 2012.

ABI Research, Internet of Things vs. Internet of Everything: What's the difference? May 7, 2014.

Adshead, Anthony, "Data set to grow 10-fold by 2020 as internet of things takes off," ComputerWeekly.com, April 9, 2014.

Alliance for Aviation Across America, "Air traffic modernization."

B

Bradley, Joseph, Joel Barbier, and Doug Handler, Embracing the internet of everything to capture your share of $14.4 trillion, Cisco, 2013.

Brynjolfsson, Erik and Lorin M. Hitt, "Beyond the productivity paradox," Communications of the ACM, volume 41, issue 8, August 1998.

Brynjolfsson, Erik, Lorin Hitt, and Heekyung Kim, "Strength in numbers: How does data-driven decisionmaking affect firm performance?" Social Science Review Network, April 2011.

Bughin, Jacques, Michael Chui, and James Manyika, "Ten IT-enabled business trends for the decade ahead," McKinsey Quarterly, May 2013.

C

Caterpillar Global Mining, "Automation keeping underground workers safe at LKAB Malmberget Mine," Viewpoint: Perspectives on Modern Mining, issue 3, 2008.

Chui, Michael, Markus Löffler, and Roger Roberts, "The Internet of Things," McKinsey Quarterly, March 2010.

City of Chicago, Office of the City Clerk, 2015 budget overview, 2015.

Consumer Electronics Association, "Record-breaking year ahead: CEA reports industry revenues to reach all-time high of $223.2 billion in 2015," press release, January 6, 2015.

D

Deutsche Bank, The Internet of Things: Not quite the Jetsons yet, but places to look, May 6, 2014.

E

Eaton Corporation, Blackout tracker annual report for 2013, March 31, 2014.

The Economist Intelligence Unit, The Internet of Things business index: a quiet revolution gathers pace, October 29, 2013.

Evans, Peter C., and Marco Annunziata, Industrial Internet: Pushing the boundaries of minds and machines, General Electric, November 26, 2012.

F

Federal Trade Commission, Internet of Things: Privacy & security in a connected world, FTC staff report, January 2015.

G

Gartner Inc., Forecast: The Internet of Things, Worldwide 2013, December 2013.

Gartner, "Gartner says 4.9 billion connected 'things' will be in use in 2015," press release, November 11, 2014.

General Aviation Manufacturers Association, 2014 general aviation statistical databook & 2015 industry outlook, 2014.

Goldman Sachs, The Internet of Things: Making sense of the next mega-trend, September 3, 2014.

H

Henderson, Catherine, Martin Knapp, José-Luis Fernández, Jennifer Beecham, Shashivadan P. Hirani, Martin Cartwright, et al., "Cost effectiveness of telehealth for patients with long term conditions (Whole Systems Demonstrator telehealth questionnaire study): Nested economic evaluation in a pragmatic, cluster randomised controlled trial," The BMJ, volume 346, issue 7902, March 2013.

I

Identity Theft Resource Center, Identity Theft Resource Center breach report hits record high in 2014, January 12, 2015.

International Data Corporation, Worldwide Internet of Things spending by industry sector 2014–2018 forecast, July 2014.

K

Kwatra, Sameer, and Chiara Essig, The promise and potential of comprehensive commercial building retrofit programs, American Council for an Energy-Efficient Economy, May 2014.

L

Lomax, Tim, David Schrank, Shawn Turner, and Richard Margiotta, Selecting travel reliability measures, Texas Transportation Institute, Texas A&M University, May 2003.

Lopez Research, Building smarter manufacturing with the Internet of Things (IoT), part two, January 2014.

M

Matheson, Rob, "Moneyball for business," MIT News, November 14, 2014.

McKinsey & Company, Industry 4.0: How to navigate digitization of the manufacturing sector, April 2015.

McKinsey Global Institute, Big data: The next frontier for innovation, competition, and productivity, May 2011.

McKinsey Global Institute, Disruptive technologies: Advances that will transform life, business, and the global economy, May 2013.

McKinsey Global Institute, Infrastructure productivity: How to save $1 trillion a year, January 2013.

McKinsey Global Institute, Internet matters: The Net's sweeping impact on growth, jobs, and prosperity, May 2011.

McKinsey Global Institute, Open data: Unlocking innovation and performance with liquid information, October 2013.

McKinsey Global Institute, Urban world: Cities and the rise of the consuming class, June 2012.

McKinsey Global Institute, McKinsey Center for Government, and McKinsey Business Technology Office, Open data: Unlocking innovation and performance with liquid information, October 2013.

Merchant, Brian, "With a trillion sensors, the Internet of Things would be the 'biggest business in the history of electronics,'" Motherboard.vice.com, October 29, 2013.

Miller, Stephen, Base salary rise of 3% forecast for 2015, Society for Human Resource Management, July 9, 2014.

Morgan Stanley, The "Internet of Things" is now: Connecting the real economy, April 3, 2014.

Morgan Stanley, Wearable devices: The "Internet of Things" becomes personal, November 19, 2014.

N

New York City Department of Health and Mental Hygiene, New York City trends in air pollution and its health consequences, September 26, 2013.

New York State Department of Transportation, State fiscal year 2012–13 annual report, bridge management and inspection programs.

Nichols, Chris, "How the Internet of Things is changing banking," Center State Bank blog, September 16, 2014.

O

O'Donnell, Erin, "Cheating the reaper," Harvard Magazine, March–April, 2013.

P

Piore, Adam, "Designing a Happier Workplace," Discover, March 2014.

Porter, Michael E. and James E. Heppelmann, "How smart, connected products are transforming competition," Harvard Business Review, November 2014.

R

Ray, Tiffany, "Fitness spending still fast despite economy," The Press-Enterprise, July 6, 2012.

Rio Tinto, 2014 annual report: Delivering sustainable shareholder returns, 2015.

S

Salas, Maribel, M., Dyfrig Hughes, Alvaro Zuluaga, Kawitha Vardeva, and Maximilian Lebmeier, "Costs of medication nonadherence in patients with diabetes mellitus: A systematic review and critical analysis of the literature," Value in Health, volume 12, number 6, November 2009.

Schwartz, Tony, "Relax! You'll be more productive," The New York Times, February 9, 2013.

Scroxton, Alex, "How the internet of things could transform Britain's railways," ComputerWeekly.com, August 2014.

Sofge, Eric, "Google's A.I. is training itself to count calories in food photos," Popular Science, May 29, 2015.

Sullivan, Bob, and Hugh Thompson, "Brain, interrupted," The New York Times, May 3, 2013.

Synapse Energy Economics, 2013 carbon dioxide price forecast, November 2013.

T

Teu, Alex, "Cloud storage is eating the world alive," TechCrunch, August 20, 2014.

Tweed, Katherine, "Entergy tests AMI voltage optimization," Greentech Media, January 27, 2012.

U

Ubl, Stephen J., "Public policy implications for using remote monitoring technology to treat diabetes," Journal of Diabetes Science and Technology, volume 1, number 3, May 2007.

UK Department for Business Innovation & Skills, The smart city market: Opportunities for the UK, BIS research paper number 136, October 9, 2013.

UK National Institute for Cardiovascular Outcomes Research, National heart failure audit, April 2012–March 2013, November 2013.

United Nations Department of Economic and Social Affairs, World urbanization prospects, the 2014 revision, 2014.

US Bureau of Labor Statistics, American time use survey 2013, June 2014.

US Central Intelligence Agency, The world factbook.

US Department of Health and Human Services, Patient Provider Telehealth Network: Using telehealth to improve chronic disease management, June 2012.

US Department of Transportation, National Highway Traffic Safety Administration, Traffic Safety Facts 2012, 2012.

US Energy Information Administration, Electric power sales, revenue, and energy efficiency Form EIA-861 detailed data files, February 19, 2015.

V

Van Marle, Gavin, "How 'augmented reality eyewear' can give you a smarter view of future logistics," The Loadstar, November 26, 2014.

Varian, Hal, "Kaizen, that continuous improvement strategy, finds its ideal environment," New York Times, February 8, 2007.

Vrijhoef, Ruben, and Lauri Koskela, "The four roles of supply chain management in construction," European Journal of Purchasing and Supply Management, volume 6, number 3, January 2000.

W

Weber, Lauren, "Go ahead, hit the snooze button," The Wall Street Journal, January 23, 2013.

Wilson, Marianne, "Study: Shrink costs U.S. retailers $42 billion; employee theft tops shoplifting," Chain Store Age, November 6, 2014.

Wojnarowicz, Krystyna, "Industrial Internet of Things in the maritime industry," Black Duck Software, February 11, 2015.